P9-DFJ-608

# How to Talk to Your Animals

# How to Talk to Your Animals

Jean Craighead George

*Illustrations by the Author*

HARCOURT BRACE JOVANOVICH, PUBLISHERS
*San Diego New York London*

Library of Congress Cataloging in Publication Data
George, Jean Craighead, 1919–
How to talk to your animals.
1. Human-animal communication. 2. Animal communication.
3. Domestic animals—Behavior. I. Title.
QL776.G45 1985 636.08'87 85-5476
ISBN 0-15-142200-1

Designed by Francesca M. Smith
Printed in the United States of America

First edition

A B C D E

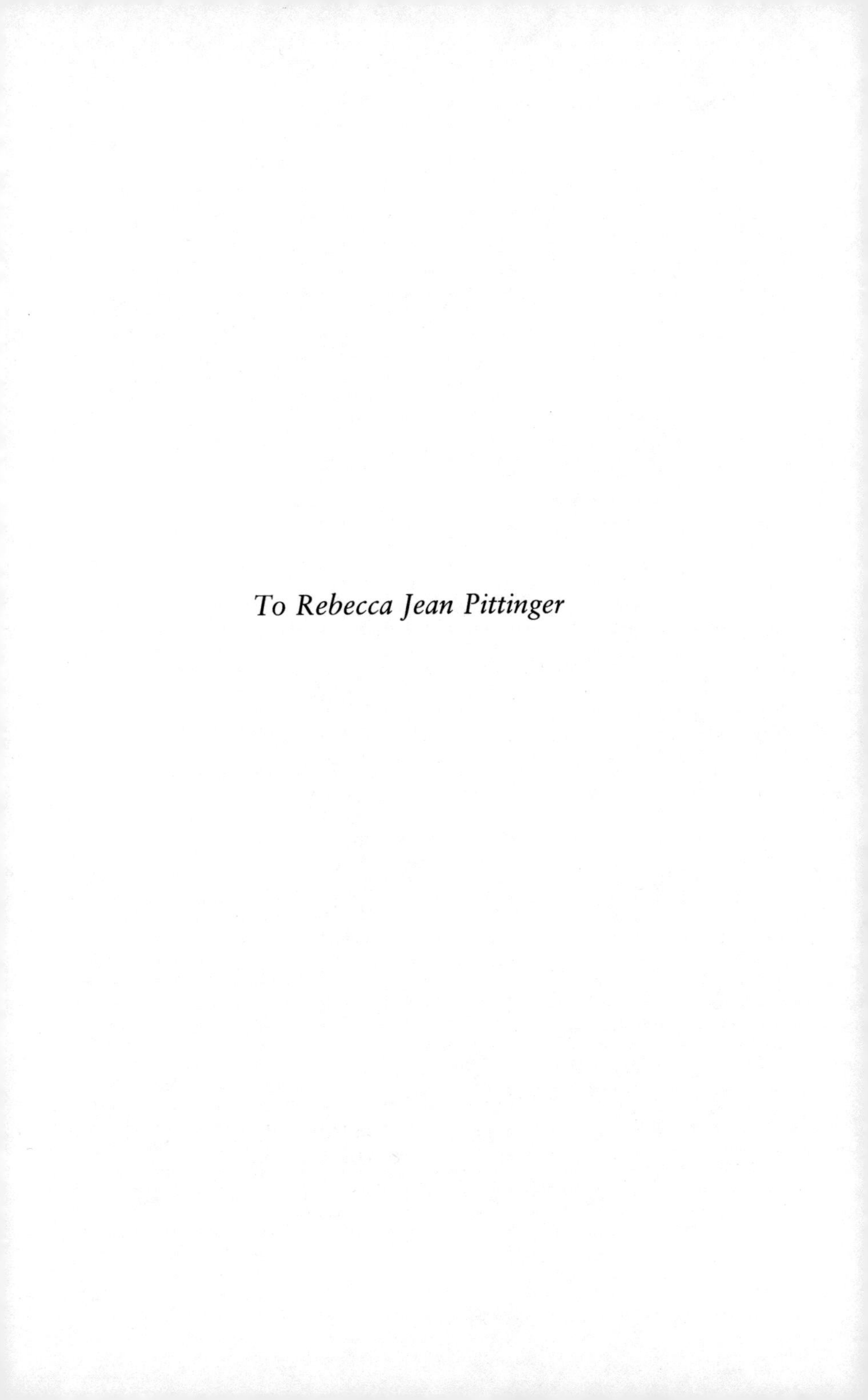

*To Rebecca Jean Pittinger*

# Contents

# How to Talk to Your Animals

# Talkier Than We Knew

A childhood hero of mine was a long-shanked blacksmith with the body of a badger and the eyes of a deer who could talk to the animals. Will Cramer was known as "the man who knows what the animals say." Chickadees clustered around his door; stray dogs came to the gate of his clapboard home as if they had heard through some mysterious grapevine that he would take care of them. Will lived in the valley of the Yellow Breeches Creek in Pennsylvania, down the road from my grandfather's house where my parents, brothers, and I spent our summers. Evenings, my brothers and I often sat on the porch railing of the country store and listened to the farmers talk about crops, the weather, and, now and then, about Will Cramer. "Will listened to a cow bawl today and told Jim Hucklefinger that she wanted to be moved to another stanchion." On another occasion Will heard a dog whine and told his owner, "Your hound wants his collar loosened."

The farmers did not quite believe Will, but did not discredit him either. They were all aware that some kind of communication went on between themselves and their chickens, cows, pigs, horses, cats, and dogs. If it wasn't exactly "talk," it was something akin to it. When the dog barked, they got up and went to see who was approaching the house, and when the cow bawled, they milked her. Communication is, after all, an action by one individual that alters the behavior of another, no matter how humble the creature or how strange the language. A toad excretes a noxious fluid when picked up by bird, beast, or man, that says quite unneatly, "Drop me." Most do.

Will Cramer not only altered his own behavior when the animals spoke to him, he altered theirs by speaking to them in their own language. He asked his dog Nick, who was standing at ease beside him one day, to play by getting down on all fours and spanking the ground with outstretched arms, as dogs do when they are sparking another dog to rough and tumble. He told the cow he wanted her in the meadow by uttering bovine sounds to attract her attention, then stepping in front and walking. In cow language, the animal in the lead is saying, "Follow me."

Anyone watching Will with animals would believe he had some gift not given to other persons, but he told me the year before he died that he just watched what the animals were doing or listened to their vocalizations, then observed what happened next. In other words, Will studied cause and effect, which is how all animal communication functions. A wolf leaves a scent of its presence and social status in its urine at the edge of its territory, and another wolf reads who it is with thoughtful inhalations and either turns away or, recognizing a friend, comes on. "Hello," we say to a person and wait to see the effect of our greeting before going on with the conversation or taking our leave.

Our remarkable communication system—our spoken language, with its infinite combinations of sounds strung together by rules—is so advanced, we believe, compared with nonhuman systems that most of us fail to recognize seemingly simpler dialogues. They are all around us: the cecropia moth calling in chemicals, the spider receiving and sending telegraph messages along a thread, the spreading tail of a peacock speaking of masculinity through a vision of beauty, the wolves keeping their space around them by howling in concert. The odors, poses, movements, displays, and the clicks, hisses, chirps, and bellows are communications. Animals, including us, speak in the four media of scent, touch, sight, and sound. Some messages use but one medium, some all four, but all media, all messages, are the self reaching out to be known by others.

Anyone with even the slightest acquaintance with animals recognizes a kitten's mew as a plea and a dog's bark as a warning. But they are only the tip of the iceberg. Animal communication has turned out to be far more complex than we had guessed until recently.

1920 was a landmark year in our awakening to the true meaning of the prettiest of all animal sounds in nature. That year British businessman Elliot Howard published his discovery that a bird's song is not the outburst of joy we in our innocence had supposed it to be, but a rather businesslike announcement specifically addressed to others of its kind. Birds sing to announce property lines, advertise for a mate, and proclaim ownership of a good habitat for the rearing and feeding of young.

Howard spent years wandering his estate, observing and charting the behavior of resident songbirds. Gradually his notes and diagrams took shape. A certain male bird was always in the same area, he sang from the same bushes and trees, and, no matter how long Howard pressed him, he never left that piece of land. Like himself, the bird had territory, and like himself, he defended it, not with words, laws, and guns, but with song. The seemingly pleasant little bird was threatening his neighbor with, "Keep off my property."

Once Howard recognized this, everyone saw it—and with a deep sense of shock, not so much because birdsong was tough talk, but because it took so many millennia for intelligent mankind to recognize something so obvious.

The discovery of territorial behavior in birds hatched a host of scientific disciplines: ethology, sociobiology, animal behaviorism, cognitive ethology—a potpourri of names for fields in which men and women have put aside the study of conventional physical zoology to observe the thinking, communications, and social behavior of nonhuman animals. Now six decades of their observations have caused a flip-flop in our thinking. We can no longer speak of "dumb" animals. Most natural scientists believe today that the birds and beasts are not automatons, performing by instinct without thought or feeling, but creatures of intelligence and sensitivity that communicate in many ingenious ways. Will Cramer was right: the animals do talk, not only to each other but to us if we listen.

"Talk" and "listen" are impoverished words to use to mean communication; perhaps our reliance on speech explains why we spent so long noticing so little animal palaver. For instance, the oldest, and still most fundamental, medium of communication is chemicals. We are aware of molecule messages as the very few tastes and relatively strong odors we can discern. We consciously use sweets and perfumes in courting. But that's nothing compared with the chemical languages we are not aware of. Every cell of our body is in constant communication through the chemicals called hormones and neurotransmitters. One cell emits molecules of insulin, and another, receiving the word, knows to take up sugar from the bloodstream. Chemical language is most likely all life's mother tongue, for bacteria, from their position on the lowest rung of the phylogenetic ladder, speak in insulin and estrogen. Even plants talk chemically. The alder tree when attacked by insects sends out a chemical vapor that "tells" other nearby alders that insects are attacking. They respond by depleting their leaves of nutrients and loading them with insect-killing toxins. Bacteria speak through chemicals that keep them together where there is food and warmth and bring them together for rare matings.

Scientists are only beginning to unravel chemical communication, but they are sure the lexicon includes the most essential distinction: self and other. Bacteria string along only with their own kind, each joining with others most like itself. On the other hand, a bacterium couples for sexual purposes only with an individual of its own kind that is not identical to itself. The first message might therefore be translated as "I am I," an announcement of species and individual self. Trading such messages, some organisms can apparently gauge their relatedness.

Andy Blaustein, at Oregon State University, recently demonstrated in laboratory tests that pollywogs recognize kin. A pollywog raised separately from its brothers and sisters, and then introduced to them and to unrelated pollywogs of the same species, chose to pal around with its own family.

We ourselves do not seem able to sniff out our degree of kinship, at least we are not aware of it, but apparently we give off a family smell. Nele, a great hairy dog belonging to my friend Sara Stein and very much a part of her family, ran toward a strange man who had parked in the driveway, tasting the air with her nose. Sara was surprised that the dog not only failed to bark at the stranger but bounded up to him in joyful greeting, tail wagging. She looked closer. The man was not a stranger but her cousin, a cousin that Nele had never met but who apparently reeked of relative.

Although our chemical reception is poor compared with a dog's, we do receive and react to some odors, often unconsciously. The smell of a lover lingering on clothes when the lover is away fills us with longing, and the scent of a baby's head elicits tenderness. We no doubt send out odors that announce our sex, but our discernment is nearly subliminal, so we heighten the difference with "feminine" perfume or "masculine" shaving lotion. Few other animals need that hype. Invisible, silent, and unnoticed as these chemicals are to us, our animals easily sniff them out.

A pet red fox I raised had been injured by a man when she was taken from her family den, and ever after she bit or ran from all men. Women she loved, and said so by leaping softly into their lap and draping herself affectionately over their shoulders. It was important to her that she made no mistake, and I soon realized by her reaction of fear even toward objects that men had touched that the red fox could discern maleness and femaleness in the human species far faster and more accurately than I, using my own cultural clues to sex—clothing, voice, and hair. While I often ponder or make mistakes through my eyes and ears, the red fox's nose was instantly correct.

Chemical communication is probably behind dogs' almost eerie ability to read our moods and feelings. While camping with me and my son one summer my son's friend Seaver Jones thought he heard a grizzly bear snapping sticks as it came toward our camp. There was no bear, but Seaver's fear was so great that Jake, his dog, smelled it and raised the hair on his back. Finding nothing himself to fear on the wind, Jake licked Seaver's face and went back to sleep.

Seaver, of course, had no difficulty interpreting Jake's affectionate face lick.

Touch speaks forcefully and intimately, especially among mammals, whose strokes, pats, nibbles, licks, and kisses share a common heritage of motherly love. Mammal mothers and babies literally keep in touch, and so do lovers. But even turtles touch. Last June I watched a male box turtle in my woods approach a female at sundown and after a few surprisingly agile maneuvers for a rock-hard creature, climb upon her back and caress her with loving taps of his lower bony carapace. His touch spoke clearly. She mated with him.

We humans speak of love as we caress, of fear as we cling, of restraint as we grip, and of aggression as we shove. We evoke laughter by tickling. We also send messages to our animals through touch, conveying thoughts that perhaps we do not consciously realize we are saying. The pat on the head we give our animals is a substitute for "Blessings on you, little man," a condescending but reassuring gesture that says, "Remain a child. I'll take care of you."

And our animals speak back to us through touch, although they are not always saying exactly what we think they are. Cats' rubbing and dogs' licking are certainly affectionate statements, but the cat rubs against us to claim us as its beloved possession, and the dog licks us to express its lovingly subordinate position.

I go to the garden to touch talk with a spider, the wizard of tactile communication. I pull very gently on its web; the owner, a female spider, pulls back on the thread. Through her telegraph system she asks, "Who are you?" I could be a male spider or an insect to eat. I pull again, and this time I say with my clumsiness, "Human being." She replies that she knows I am a predator by running up another thread and hiding.

I try other things on her. I shake the web with a slim stalk of grass, hoping to say, "Food," then I sprinkle water to say, "The dew of dawn." I wait for her reply. She does not come out. I toss a katydid in her web, and this time she answers me by dashing out and killing it. I learn her definitions of the insects, for she reacts differently to grasshoppers and gnats, enshrouding each with a defining silk that virtually bears the insect's name. A thick band thrown from a distance is "bee," a broad swath wrapped at close quarters while her feet spin the prey reads "grasshopper," a single thread says "gnat."

To a spider, touch brings messages from a distance along the far-flung web of her communication system. Were a male spider to court her—by drumming his love tattoo from a safe distance before approaching his dangerous bride—she would receive the message in time to avoid harming him. But touch ordinarily requires actual contact, and smell, too, is commonly an intimate communication. The prime long-distance sense in higher land animals is vision. We see and interpret before we encounter. Visual signals, such as a stallion's proud posing on the rim of a distant hill, can be sent over greater distances than animal sound—miles compared to half a mile for the loudest noisemaker, the cicada.

Strangely, until very recently we remained unaware of visual signals of our own more subtle than a wave or a smile. Now, with our knowledge of "body language," we are in a better position to appreciate the expressive faces, poses, and gestures of other animals. Even bodily proportions speak. The proportions of kittens and puppies elicit a parenting response in us, and we say, "Aaaaw," and pick them up. What is talking to us are the large head and eyes, the small body, and short, plump limbs. "I am cute. I am helpless, hold me, care for me," the baby proportions cry out. We are not taught the message of proportions; it comes from deep in our nonintellectual being, down with loving and needing.

Our animals also reply to babyish proportions with parental behavior, even when the baby is another species. Our bluetick hound, Delilah, had such a response to my son Luke when he was a baby. She licked and encircled him and leaped into his crib to curl warmly beside him when I was in another room. She did not do this to Luke's six-year-old sister or four-year-old brother or to me. That's baby power for you.

Adults display their power by posture, and these visual signals, too, are well-nigh universal. The signal is given by making oneself appear big. The wolf says "I am leader of the pack" by holding his head higher than the others. The gull chief throws out his chest and stretches his neck above his kin to announce his dominance. The male songbird puffs up his feathers to speak of his importance and aggressiveness. Kings, queens, and presidents are taught to carry themselves erect. Military officers thrust out their chests like pigeons to say the same thing—"I am dominant and powerful."

The opposite message—one that indicates a low position on the social ladder or an acknowledgment of inferiority in an encounter—is given by making oneself appear small or childlike. The wolf and dog roll over puppylike on their backs in deference, the low-order gull pulls in its neck and crouches, and the humble servant bows to royalty. These postures are extremes and therefore are easy to recognize.

But there are as well an infinite number of other communicative poses, including those of the face, that amplify, dampen, and even contradict "outspoken" messages. Many of our facial expressions—from smiles to frowns—are congruent with what we are saying: when a dear friend says, "I am so happy to see you," the broad smile amplifies the spoken message. We are aware that this is not always so. Even children pick up an insincere smile, despite its being coupled with the words "What an adorable child!" By filming people in stressful situations, such as getting fired, psychologists in California have compiled a "dictionary" of facial messages that contradict spoken statements. Written on these faces are such messages as "masked anger," "overly polite," and "resigned compliance." Facial expressions speak truer than words.

Few other animals have such well-developed—or even as many—facial muscles as we, so a horse's or cat's face seems rather blank to us. To understand what they are saying, we have to learn to read ears, whiskers, the pupils of the eye, and the glint of teeth. They, too, can modify a voiced message. A dog growling with ears pricked is quite sincere in its threat to bite. A dog growling with ears lowered is as frightened as it is annoyed.

Light is another form of visual communication in the nonhuman world, and none uses it so beautifully as the firefly. I grew up in firefly country. When I asked my grandfather why they had lights, he answered, "Because lights are lovely. That's reason enough," and I accepted that. In the late 1950s a persistent firefly observer discovered something that changed the summer night for me. "The lights are a language," he told me one evening. "A simple code. When flashed, the signal is 'Yes'; when off, it means 'No.' He pointed to the lights above the meadow.

"The fireflies that are blinking and climbing upward are males advertising their sex, which species they are, and their sexual readiness to the females in the grass."

Now as I walk along a stream at dusk, I see loveliness and much more. I see the drama of firefly love. A male glows at the top of an ironweed, lifts his hard outer wings, vibrates his gossamer underwings, and takes off. He climbs, flashing his lamp at a leisurely beat as he courts the females first with slow strokes, then faster and faster. They watch from below. When he reaches the tops of the trees, his call comes even more urgently; the night is passing. A female at the base of a clump of foxtail grass is finally aroused. Her light goes on—"Yes." The male fastens his many-faceted eyes on her and descends. He, too, is ready. His light says, "Yes."

The lamp in the grass goes out—"No." Did another male find her and turn off her light? The downward-gliding male still flashes his bold visual cries and is answered by another light in the grass. He descends, flying past a male who at that instant is caught by a bat, nipped, and crushed with his light on. He falls, his continuous glow becoming a statement of death. It is not a communication, for no firefly upon seeing it changes its behavior. All continue to shine and signal. The glowing body falls to earth like a star and goes out. "No."

The male I am watching drifts into the meadow grass. The "No" female of the foxtail lights up again—"Yes." He touches down on her grass blade.

Both lights go out. The firefly embrace is invisible to bat, frog, skunk, and me, and a conversation in lights is over.

As lovely as this conversation is, to most people the ultimate in visual communication is the peacock. Every pattern and color, from the iridescent head, chest, neck, and golden coverts to the spreading tail and wings, says, "Choose me. I'm beautiful." The peacock's beauty is a result of what is called sexual selection. The peahens have created this wondrous creature over the eons by selecting the most beautiful male to be father of their offspring. And because of their eye, we have among us the super star of visual talk.

Honeybees employ dance to communicate visually with their sisters. Sociobiologist Edward O. Wilson compares their language to our own because they can, like us, speak in symbols. Through a dance one bee passes on to another information removed in time and space.

A forager bee returning to the hive after finding a good source of nectar and pollen alights on a hive wall and runs a figure eight until she gathers her sisters around her. As she dances she puts the emphasis on the *straight-line run* down the middle of the figure by vibrating her body. Were she dancing on a flat surface, the straight-line run would point directly to the flowers. But the hive wall is vertical, so the bee, in effect, pins the map up on the wall. Just as we might use the compass points to describe direction—twenty degrees east of north—she uses the present position of the sun—twenty degrees to the right of the sun. But the hive is dark, so, by convention, she uses "up" to indicate the sun's position. We use a similar convention: north is "up" on our maps.

The distance to the flowers is calculated and translated into time. In one species of honeybee, a run of one second duration indicates the food is five hundred meters away. A two-second run states that the nectar is two kilometers from the hive. During the dance the sister bees call upon other communication systems. They touch the dancer to discern the vigor of her waggling, an indication of how promising she considers her find. The smell and taste of the pollen and nectar on the dancer tell them what flowers they will be looking for. Once briefed, they take off; and the majority arrive dead on target—which I couldn't do by following verbal directions.

Communications so unlike our own as the bee's dance have taken researchers untold hours of observation and experiment to decipher. They are not languages that Will Cramer, blessed as he was with intuition, could have translated. He, like most of us, was more at home with sounds.

As an admirer of animal sonance, I seek out, in the early summer, the most famous theater for croaks, growls, squawks, songs, trills, pipes, and whistles—the swamp. Up from the water surface, out of the reeds, and over the floating plants soars the full choir of animal voices. Insects, amphibians, reptiles, birds, and mammals speak with such volume that I am tempted to add a human song to the glorious clangor. One spring night while camping in the Okefenokee Swamp of Georgia, I heard between sundown and sunup barred owls hooting, alligators bellowing, a wildcat caterwauling, raccoons growling, twenty-two species of frogs piping, and several hundred species of insects buzzing, clicking, whining, and humming. In the dawn light, the shriek of the wood ibis, the carol of the vireo, the rasp of the pileated woodpecker were added, one voice upon another, hundreds of species strong. It was a night and dawn of expressive doings, and my great pleasure was that I could translate a few of their phrases because of the new knowledge in animal communications.

That new knowledge is, however, minuscule. We are on one side of a great abyss that separates human from animal language, and the distance often seems uncrossable. How can we really know what is going through a dog's mind when it wags its tail? We can never get inside its brain. But we cannot get inside each other's minds either. How do I know you see the color green as I see green? What is important is that the dog sees something and is wagging its tail. What it is that makes the dog happy is where we should be looking. As it happens, dogs wag their tails only to living objects—butterflies or people—but never to anything inanimate or dead, not even to a bone. That says something about dogs, even if we can't come up with an exact translation for what they are trying to convey to a butterfly. There is a why to smell, touch, sight, and sound signals, and the why is to be found, as Will Cramer saw, in what happens next. When dogs wag their tails to one another, they then approach in a friendly manner.

So we know that friendliness spells happiness to dogs. Perhaps they invite butterflies to play across an abyss much greater than the one that separates our two mammalian species—one that never will be bridged.

When I read who on this earth was the first speaker to have organs specifically designed to make noise, I felt as if someone had thrown a rope across even that impossible abyss for me. The first voice was so simple and so sensible. It was uttered by an insect, a grasshopper, a fragment of which lies trapped in a fossil 200 million years old. The grasshopper had sound-making files on its wings almost like those of the modern grasshopper. Back in that mist of time, the file owner rubbed his wings together and vibrated the air. The ripple reached a receptor on another of his kind, and she followed the sound to its source. They mated. The first noisemaker needed never again to waste time and energy running around looking for a mate.

To become aware of the languages other species speak is thrilling, but to exchange conversations in these lexicons with another species is the quintessential effort. We all wish we could hold a discourse with our animals, if only to know how they like us. A few people manage to communicate across the abyss with little effort: the rare Will Cramers, a handful of scientists. But there are universal messages that wing across the gap when we least expect it. My brother, John, had a pet coyote named Crooner many years ago in Wyoming. The pup moved on the fringes of John's family life but was never really a part of it. He would come in to play, request food and affection from time to time, and then vanish into his own world in the sagebrush. One day while John was nailing shingles on the roof of his house, he saw his two-year-old son tumble into the irrigation ditch. He still does not know if he yelled or simply scrambled frantically to get off the roof, but the coyote received a distress message from him, appeared like a white knight, and pulled Derek out of the water.

"Danger to the young," he must have conveyed, and the coyote understood it.

Will Cramer never opened a book on animal behavior or took a course in dog obedience, yet he talked to the animals, according to his own analysis, by listening for those expressions we all have in common and conversing on those subjects.

One day I asked Will to tell me how I could let his cat know I wanted to be her friend. He demonstrated by making soft squeaky noises with his lips and holding out his hand. I made soft squeaky noises and held out my hand—and was swatted.

"Too much eye," Will said. "To ask a cat to be a friend, you have to mostly look away." Then he added, "Besides, you smell wrong. The cat smells that you think people are the only critters who can talk."

And he was right. An inside voice from my background in a scientific family was pulling, saying, "Don't anthropomorphize. Don't attribute human thoughts and emotions to animals."

Years later, after raising 173 wild and 50 domestic pets, I still felt that pull, although I had learned to peep like a quail and make hand signals to our mink to invite her to leap and tumble on the back of the couch. Despite such enormous triumphs, my human viewpoint prevailed—only people can truly converse.

It was not until I read a scientific paper on the language of the wolves by the esteemed animal behaviorist Rudolf Schenkel of Switzerland that I believed there were other languages and perfectly brilliant conversations going on on the other side of the abyss. Wolves talk among themselves, Schenkel observed, with eyes, postures, scents, body contact, mood, and voice. And we can use the same signals to talk to wolves.

A few phone calls to wolf enthusiasts informed me that wolf talk and human talk was being exchanged at the Naval Arctic Research Lab in Barrow, Alaska, on the North Slope where the Arctic Ocean laps the land. I approached a national magazine, took on the assignment to "talk to a wolf," packed up, and flew with my youngest son, Luke, to the Top of the World. In that Arctic wildland I interviewed the authorities on wolf behavior and observed the wolves for many long twenty-four-hours-of-daylight days, listening for the only thing I understood at the time—the sounds of whimpers, howls, sniffs, and barks. After practicing the sounds, I felt I was ready to make the crossing of the abyss. I selected a beautiful female I called Silver. She was almost white, her eyes were gold and serene, her disposition, despite her confinement in a large enclosure, was sweet. In broken wolfese, I whimpered my affection and asked her to be my friend.

She ignored me.

"Your pose is wrong," wolf-language expert Dr. Michael Fox said. "Say it with your body." I stood more confidently and moved with more dignity. Still no response. Fox assumed a kingly attitude. I copied his posture and lifted my head with authority, but my whimpers and poses still brought no response from Silver. I was not getting through. There was no change in her behavior. She walked past me as if I were less than a pebble on the tundra.

The day before I was to leave Barrow, I went back to my notes and read something I had skimmed over. The wolf father leads the pack on the hunt, the mother attends to the pups, and baby-sitters and the puppies mock fight, tumble, and play games. The wolf family is very much like the human family; within it conflicts arise and arguments are settled, jobs are assigned and done.

A feeling of similarity replaced my feeling of strangeness. Silver and I had much in common. We were both parents, females; we both watched and disciplined our young, taught them safety rules, encouraged their play, and stopped their aggressions. We had common ground on which to meet and talk. I put away my notebook and went back to the animal lab.

When I approached Silver's enclosure on this last visit and whimpered to her, she hesitated and . . . passed me by.

But as I turned to leave, Silver stopped in her tracks, then galloped back to me. She smiled. Wolves pull back their lips and smile just as we pull back our lips and smile, and for the same reasons. They are pleased, or they like you.

With that smile she let me in. I was not a pebble on the tundra. I was in her consciousness. Having admitted me, she continued the conversation. She wagged her tail, looked me in the eye, and whimpered in a high thin voice, "Uummum," the wolf plea for friendship.

I whimpered, "Uummum." Silver wagged her tail enthusiastically, spanked the ground with her forepaws, and asked me to play. I, at long last, was talking to a wolf.

I did not do anything differently that last meeting with Silver, but I felt differently. I had drawn on our likenesses, not our differences, and, in doing so, I had come down from the pedestal that human beings put themselves on. As Will Cramer might have said, I "smelled right."

The conversations that await us as we learn to reach across the abyss are no doubt full of surprises. One man, Dr. George Archibald, founder of the International Crane Foundation, Baraboo, Wisconsin, must know this better than anyone. He heads an institution dedicated to saving from extinction the stately cranes, among them the whooping crane. Archibald has studied crane communications and their behavior so thoroughly that he can not only share the universal meeting ground with them but *really* talk and get them to answer.

A female whooping crane named Tex, one of the last 150, was given to Archibald by the San Diego Zoo. The bird was imprinted on people; that is, she had been raised from an egg by humans, and she thought she was human, too. When she matured, she fell in love with Archibald.

At five years, when she was of breeding age, she would not mate with her own kind for love of Archibald, and so she was artificially inseminated.

It did not take. Tex was unable to ovulate without the stimulation of courtship. But she, like others raised in captivity, would not tolerate the attention of a bird of her species.

Archibald saw a possibility in his relationship with Tex and decided to try his luck.

He would walk slowly into her outdoor enclosure and begin to speak to her of love in her own language. When she flapped her wings—the equivalent of a wink from a lady in our own body language—he would flap his preposterous and awkward arms. As clumsy as this suitor was, she encouraged him because she loved him and he was talking crane talk. She would float into the air and drop back to earth. He would jump and come down. He would utter strange cries in response to hers and spend as much time as he could spare walking and waltzing with Tex. She was inseminated and laid an egg. But the egg was not fertile. Tex had not reached the climax of crane courtship, and without achieving this, her eggs could not be fertilized.

The following breeding season Archibald took time off from his travels and duties and devoted the day to Tex from sunrise to sunset. They met in the morning and flapped arms and wings. They leaped lightly over the grass and jumped into the air, Archibald stretching his neck in imitation of Tex's stretch. They walked together, gathered worms together, and finally built a nest of grass and corncobs together. One morning when Tex danced with more excitement than Archibald had seen before, she was artificially inseminated. It took. A fertile egg was laid on May 3, 1982. The whooping crane Gee Whizz was hatched on June 3.

In the following chapters you will learn the history, social behavior, and languages of dogs, cats, birds, and horses—those animals that touch our lives most closely.

"Animal talk is on the other side of words," Will Cramer once told me as he explained his uncanny communication with his dog Nick. "He watches my eyes to know what I am thinking; I watch his." But Archibald could not have courted a crane without studying its biology and learning the meanings of its whoops, leaps, and flaps.

So on the one hand, my guide will be the intuition of pet owners who, by their experience, just know how to talk to animals, without knowing exactly how or why what they do works.

On the other hand, my guide will be the scientists who have unraveled a few of the mysterious threads of animal communication through careful and objective research.

For those who need more than Will Cramer and the scientists to make them believe that they, too, can cross the abyss, may I say that if alder trees can communicate with alder trees, why not dogs with dogs, and cats with cats?

And for those who accept this but doubt that cats can talk to people and people to cats, let me mention that creosote bushes can communicate with weed and flower seeds in the soil beneath them—and tell them chemically not to grow.

When I learned that, I went out and talked to everything.

# The Dog

Classes were over for the day at the University of Alaska in Fairbanks, and Steve Wood hurried down the corridor of the zoology lab to meet and walk home with Orion, his malamute. The dog, who was curled in the snow near the entrance to the building, heard Steve's footsteps among the many, got to his feet, and waited for him at the door. When Steve appeared, Orion pulled back his lips in a dog smile, and his curly tail beat over his back like a metronome.

"Come on, fellow," Steve called, and Orion swept toward him like an onrushing avalanche. His head was slightly lowered, ears down and back. The lower lids of his eyes lifted, softening his facial expression. His tail wagged his whole rear end joyously. He sniffed his master.

"Hello," Orion was saying in dog talk.

Steve dropped to his knee, took the dog's head in his hands, and shook it affectionately. The movement was of father to son.

"Hey, Orion, how's the boy?"

Orion took a deep draft of Steve's odor and licked him under the chin to tell him he was a person of great status. Steve replied by patting his head, a friendly way of agreeing with the dog's perception of his rank. Orion lowered himself onto his belly and uncurled his magnificent tail in deference. Having made that statement, he arose and barked to bind them together after their long separation and to announce himself to Steve in a more individualistic way—"It's me. Take notice."

Steve hugged the furry malamute and looked into his eyes. Orion gazed back, and after a moment they started off across the campus, the man grinning, the dog wagging his tail—two expressions that mean the same thing.

On the way home the malamute spoke to his master again. The message made no sense to Steve, and for a few dangerous moments he did not understand its purpose.

"We started down the road side by side, walking along quietly as we do," Steve recounted to me several months later. "It was getting dark, so I stuck up my thumb to hitch a ride. A young kid driving a sports car pulled up and stopped. When I made a move to open the door, he spun the wheels and took off, rubber burning, stones flying. He disappeared in smoke. I shrugged—'The kid's nutty'—and walked on.

"A short way down the road, Orion leaped on my chest with his front feet. That's taking over in dog talk, a way of telling me he was boss. But he had never said that to me before. He barked, dropped to the ground, and jumped on me again. He's strong, and I staggered backwards.

" 'Are you crazy, too?' I shouted. 'What's the matter with you?' I yelled at him again, and, with that, he threw all his weight against me. I fell into the snowbank. It was then I saw the sports car coming down the road wide open, headed right at the spot I had been a second before."

**TWO OF A KIND**

That there is communication between dog and man no owner will deny, but how to read it is not always clear. Dog owners the world over perform the ritual greeting Steve and Orion exchanged every day, aware of the mutual happiness it expresses but perhaps not aware that they have also discussed rank, relationship, obligation, and individuality. Steve Wood's perceptions had been sharpened by his years of studying biology, but even he was surprised at the rescue.

"Apparently Orion had sensed something crazy about that kid," Steve said. "And when he heard the car turn around and start back, far beyond my hearing range, he became my boss and ordered me off the road. He saved my life. I'm convinced of that."

We have poor senses of smell and hearing compared with dogs, and their communications among their own kind depend largely on both. Our precisely muscled faces show emotions more dramatically than dogs' faces, and so we sometimes fail to notice their subtler expressions. Lacking a tail language ourselves, we see only the "loudest" tail words and miss the softly spoken statements.

Nevertheless, the two species have much in common. Dogs and people both speak about rank, territory, and relationship. To both, most communication is in the service of maintaining social order and the individual affections that glue the group together. To discuss these things with dogs, it is important to know their social structure and how they converse about it among themselves. Dogs talk to dogs the same way dogs talk to us. Once you know what they are saying, you can talk back to them.

The ancestors of the dog were wolves who, like early humans, had specialized as big-game hunters. In fact, wolves specialized much more than we did. The whole canid, or dog, family has a short digestive tract that can't handle much in the way of vegetables. In the wild, canids are obliged to live on meat. Although any wolf can catch a rabbit—and it will even stoop to rodents if it must—its favorite food is the large herd animals also preyed upon by humans. We and they have been, and we now are, the most important predators of hoofed animals, such as deer, elk, and caribou. These large and swift animals can outrun both a single man and a single wolf and can kill with the strike of a hatchet-sharp hoof or gore with an antler. Only the large cats, which leap on the back of their prey from ambush and hang on with their claws, can dispatch a deer without help. Wolves have evolved to hunt with their kind, as have humans.

Cooperative hunting means that individuals coordinate their actions, each one communicating its intentions, so that all the others can figure out what they must do to get the job done. Wolves and people evolved a system of coordination that relies heavily on a leader who, as we so often hear, must have "fearlessness," "initiative," "good communication skills," and that mixture of aloof discipline and reassuring goodwill we might lump together as "management ability."

The boss wolf is called the *alpha* wolf. He (more rarely she) decides which animal to pursue, orders the attack, or calls off a futile chase. He chooses which portion of their perhaps ten-square-mile territory to "camp" in temporarily, and he leads the pack to a more promising area when that one has been depleted of game. The leader is respected, indeed loved, by his pack. They look to him for guidance when the crack of a twig might mean danger. They turn to him when a raven calls an alarm or an unfamiliar scent is on the wind. They are like children looking to a parent for reaction and guidance.

There is also an alpha female who dominates the pups

and other females and who is the mate of the alpha male. On rare occasions a female will be the leader of the pack, probably because she outranked the male next in line for the job when the alpha was killed or died of old age.

Second in the hierarchy is the vice-president, or *beta*, the alpha's best friend and the animal who takes the lead when the alpha tires, who assists him and instantly responds to his wishes, and who is relied on to initiate some activities on his own. Every wolf in a pack that may number from two to more than twenty has a specific place in the hierarchy, each labeled by biologists in the order of the letters of the Greek alphabet.

The hierarchy serves both business and pleasure. It erases the uncertainty that an individual might feel if it did not know how to behave or how others would respond to its behavior. In a wolf pack, social relations are as benign as in those finely tuned human families in which each member is reassured by the reliable behavior of all the others. And, as in such a family, each member has a job to do. One may guard, another scout or track, chase, ambush, or attack. These roles are less strictly defined by social position than by talent. The alpha couple, for instance, is always prominent in the hunt, but allowances are made for age, sex, health, and other circumstances, so an individual's job may change from one hunt to another and over its lifetime. On the other hand, an individual with an unusually good nose will probably be sent out for as long as it lives to scout the odors.

## SPEAKING OF STATUS

Social position can change among wolves as it does among humans. This is something a dog owner may notice when he introduces a brash young dog into a family in which there is also an aging pet. The young one may challenge the older dog's position—chase it from its food, take over its favorite sleeping spot, push for more affection—and then the old dog's status falls. The dog is noticeably depressed: head and tail droop, the ears are held low. Like the old family dog, an older wolf may give up his position. An aged alpha in Mount McKinley National Park walked off and left his pack when challenged by a younger leader. A "lone wolf," he roamed the forest living off small game until he died several months later.

This is the permanent posture of the wolf who is at the bottom of the ladder. He or she is called the *omega* (Z) and is the weakest member of the group, the one who defers to all others. The omega is, however, an important part of the group, for that individual defines the bottom of the social hierarchy.

With each individual settled into place, from alpha to omega, peace reigns in the wolf society and the animals can work together to the benefit of all. The only time a wolf pack is disrupted is when the boss weakens or dies. Then there is a scramble for the top, and individuals behave aggressively or fearfully until order is restored within a new hierarchy.

With the evolution of the cooperative hunting society, a complex communications system developed that includes giving orders, disciplining, expressing affection, and pulling rank to get work done. Much wolf talk is about rank, each one announcing its position in scent, poses, and facial expressions or replying to the assertions of others.

The leader of the wolf pack speaks of his rank by holding his head and tail higher than the others. If any doubt his position or violate his rules, such as trying to mate when there are too many in the pack for the number of prey animals, he stands over them. They respond to his posturing by lowering

*Omega greets alpha*

themselves before him, by crouching, or by drooping heads and tails. Once his dominance is established, he need only glance at a wolf that is getting out of hand to get it to cooperate.

Because people use similar signals among themselves, dog owners intuitively raise up to tower over their dog when it is behaving badly and stare to warn it away from the pretzels on the coffee table. These messages work because, to a dog, its master is its alpha. One of the great pleasures of having a dog is, in fact, that no matter what our status is among other people, we are Number One to our pet.

The glue of a cooperative society is friendliness and respect for one another. Wolf friendliness is expressed in smiles—not the broad grin that humans are capable of, but a slight opening of the mouth and pulling back of the lips at the corners (some dogs expose the teeth like people)—and in smoothed foreheads, wagging tails, licking, and mouthing. Respect is expressed by an acceptance of the less talented

*Dog greets master*

individuals. The lowly are not abused unless food is scarce and the lives of the pups are threatened.

Friendliness and respect within the pack are in contrast to how a wolf feels about strangers. They are simply not allowed in the pack's territory; if a stranger doesn't leave promptly when chased, it will be attacked. A stranger can gain entry into the pack only when numbers are low and another pair of eyes and ears and four legs are needed. Strangers may also be admitted to a pack; they bring in a fresh genetic strain. Usually pack members are related to one another and have known each other from birth. They are a family.

This, too, is something dog owners respond to intuitively. We give a dog a "place" within our family—a bed and a bowl—and we expect it to respect and love all family members, regardless of their size, strength, or personality. You'd probably feel hurt if your dog took to strangers indiscriminately. Many people keep a dog expressly to keep their family safe from strangers.

Because wolves share with us an alpha social structure and strong family ties, and because they are even better hunters than we with their keen ears and noses, it is natural that they were the first animals to be domesticated. The oldest fossil remains of dogs have been found in the Yukon and have been carbon-dated at twenty thousand years, many millennia before chickens and sheep were brought under our stewardship. Dogs live today in closer association with man than any other beast.

Zoologists in general agree that the dog, *Canis familiaris*, is descended from tamed wolves. Taming a wolf pup is not difficult, and it is still done by canid experts, Eskimos and Aleuts. Since dogs can also breed with other close relatives—including the dingo, coyote, and jackal—the numerous breeds and mixes of dogs we have today may represent a delicatessen of canid genes. But whether you have a Pekinese, whose origins are lost in time, or a malamute, which until only decades ago was bred back to wolves to improve the stock, conversations with your pet are mostly in wolfese, and to its human alpha and family the dog brings all the love and respect the wolf gives its leader and its tribe.

Also, although domestic dogs by now have acquired, through selective breeding, a wide range of sizes and shapes—from the Irish wolfhound that stands three feet high at the shoulder to the six-inch Chihuahua and from the short-legged dachshund to the blockbuster bulldog—all behave as though they were among us to hunt.

The forty million dogs in the United States are mostly household pets, but their heritage as hunters dominates their behavior. The dachshund with its short legs will dig and go into a hole given its druthers. The English sheepdog herds its human family if it does not have sheep, using the same strategies a wolf might use to run a herd of caribou. The hound, given a run in the woods, tracks rabbits, raccoons, or—most exciting of all and a wolf favorite, too—deer. The pointer will freeze behind a city window and point the pigeons on the sill, and the Labrador with no birds to retrieve brings back rocks and shoes. Playful as these antics may seem, they are all part of the hunting repertoire of wolves at work.

## WILD CONVERSATIONS

Animal behaviorists are gaining insight into the language of the canids by studying their natural behavior in the wild or under semiwild conditions, such as those where wolf packs are enclosed within a territory two or three acres in area. Interested in learning dogs' mother tongue and fortified with background material from zoologists and animal psychologists, I journeyed to the Naval Arctic Research Lab at Barrow, Alaska, to observe semiwild wolves, a pack of seven in a quarter-acre enclosure. There Dr. Edward Folk, a physiologist with the University of Iowa, was studying the heart rates of an alpha and omega as they confronted each other alone and then within the pack. Results? When the alpha and omega faced each other alone, the alpha's heart rate was low, the omega's racing. When returned to the pack, the alpha's went up, the omega's went down as he slunk off to his low but comfortable position in the pack.

While in Barrow I learned that Dr. Gordon Haber, then a graduate student, was gathering data on the wolves of Sanctuary River in Mount McKinley National Park as a follow-up on Adolph Murie's famous study "The Wolves of Mount McKinley." Through bush pilot, radio, and word of mouth I contacted Haber; and through word of mouth, radio, and bush pilot came back the invitation to visit him. I joined him three days later.

The pack consisted of a gorgeous black alpha, his vice-president (the beta), the alpha female, the omega, and nine four-month-old pups. I watched them from a distance through a spotting scope, for wolves are man shy. At first, it was hard to pick up even the gist of a conversation, much less the nuances, but by the end of ten days' spying through the spotting scope, I could compile a typical night's communications.

They slept off and on all day. Twilight does not come to the sub-Arctic in midsummer until 11:30 P.M. In the bright six–P.M. light one evening in late July they got to their feet, yawning and stretching and glancing at the alpha to see what he was doing. The pups got up. Some had slept in shallow saucers scratched into the tundra and others in the summer den, a tunnel in the ground where they also hid when frightened. When the pups were three months old, the pack had moved to this summer camp from the whelping den, or nursery, three miles away on a high ridge that overlooked a river valley where moose and caribou lived or trekked. The present camp was out on the gray-green alpine tundra under snow-covered peaks, far from grizzly bear territory.

For the next half hour the wolves conducted a ceremony that would seem vaguely familiar to anyone who, after spending a quiet Sunday reading the paper or watching a ball game, gets up and arouses his dog to the expectation of an outing. The ceremony started quietly and built in excitement. First, the alpha female walked up to her mate and leader. He sniffed her nose, then she sniffed his. They wagged tails and stood head to tail. The alpha smelled the female's genitals briefly, just enough to check on her state of health, his "How are you?" to her. She brushed against him, and he enclosed her muzzle in his mouth, gently and with affection. This action of the alpha says, "I am the leader." Dogs have lost this particular word of their mother tongue. In their domestic dialect, they lift a paw and place it on the shoulder of the lower-ranking individual to say they are the boss.

In response to the alpha wolf's statement of rank, the female licked his face and chin to tell him that he was indeed the leader, and a wonderful one at that. This pleased him. His tail flashed back and forth. I recalled my dog licking my hands and face and realized that those licks were not kisses but statements of my alpha rank.

The next communication came from the omega, a male in this case. He approached the alpha with his ears back and pressed tightly against his head, his mouth drawn down, his tail between his legs. The alpha lifted his head, wrinkled the top of his nose slightly, and pulled up his lip on one side to

*"I'm boss."*

*"You're boss."*

expose a single sharp, white canine tooth that contrasted with the black inner lip. The omega crouched lower, until his belly touched the ground, then rolled to his back. That is the pose of utmost canine humility; the omega was stating his rank, the bottom.

The beta trotted up, keeping his head and tail slightly lower than the alpha's, higher than the omega's, but at the same level as the alpha female's. The two of them seemed, in this situation, to be on a par. The alpha sniffed the beta's nose. Among wolves, the animal who sniffs first in the greeting ceremony is dominant. That word hasn't been lost. If, out walking your pet, you come upon a strange dog, you need not worry that it will be the aggressor if your dog sniffs it first. Like the wolf, he who sniffs first is boss.

As each wolf, in turn, exchanged greetings with the alpha, excitement grew. The pups jumped and ran among their elders. They nipped their tails and, if they could jump high enough, their ears and were patiently tolerated by the adults. One pup, carried away by the festive spirit, attacked the alpha. That would have called for severe discipline had it been an adult, but the old father simply looked at it condescendingly and wagged his tail.

## GETTING DOWN TO WORK

The family greeting over, the alpha sat down, and the pack instantly gathered around him, their tails wagging furiously like pet dogs who gather around their master as he sits down to pull on his boots. The pups leaped, ran in circles, barked, bounced away and came back, acting exactly like my dog when I pick up the car keys. Something, the wolves were saying, was about to happen on the tundra along Sanctuary River.

The alpha threw back his head and howled a lugubrious note. It began low and crescendoed. The female joined in, then the beta and omega, each on his own note, each harmonizing with the others. This vocalization said pretty much what the prancing and jumping had said—"Something is going to happen"—but now to this was added, "Everyone get ready to cooperate." The howling reached a melodious climax, the pups joining in with high-pitched yipes, and, as I lay listening and watching, the treeless tundra in a bowl of mountains became a concert hall for the song of the wolves.

The howl given at this hour of the evening is called the *hunt song*, and it is sounded to coordinate the group and focus their attention on the job ahead. Dogs do not howl before a hunt, not even the foxhounds and beagles, for that song has been lost in breeding. Instead, they bark when the gun comes down from the rack or when the horses trot into the field for the same reason wolves howl their hunt song—to coordinate their activities and to psych themselves up for the hunt. Your pet may or may not bark when you pick up the car keys or put on your coat, but the other expressions are there—the wagging tail, bouncing, and running back and forth. When the ceremony was over on the tundra, the alpha led the adults over a ridge and out of sight, his fur flowing like a robe. The hunt was on.

## HOME TALK

The pups, for all their exuberant participation, were left behind in the care of the omega. Wolves are devoted parents. The mother stays in the den with the newborns for the first two to five days in zoos and about ten days in the wild. During this time the father leaves food for her at the den entrance and pauses to listen to the sounds of life in the darkness, tilting his head and wagging his tail, passing on the message to the others that the pups are all right. They, getting the message, wag their tails too. Around the fifth day the mother may leave the pups to go off to feed on a kill, but she returns promptly to her charges. Her hunting skills, more than those of other pack members, are very closely integrated with the alpha male's, making her services so valuable that when the pups are between two and three weeks old and able to keep warm without her, the mother will join the hunters. Always before she goes, she assigns a baby-sitter; wolves never leave their pups unattended.

The female of Sanctuary River assigned the job to the omega, although the omega is not always the sitter and is only rarely a male, as in this case. When he attempted to follow the hunters, she came back over the ridge, met him, and, lifting her head and tail, loomed over him to display her rank. He refused to return. Both canine teeth gleamed from beneath her pulled-back lips. Her forehead creased in a frown. The omega cowered and slunk back to the pups.

For the next twelve hours they jumped on him, bit his tail and feet, knocked him over, and, when he lay down to rest, piled on him. Pester him as they did, they listened to him. When the baby-sitter spotted a hiker coming down the riverbank, he gave a bark of warning, and the pups disappeared into the tunnel.

Around 6:00 A.M., after an entire night of abuse, the omega walked to the ridge and stood stone-still. Then he wagged his tail slowly, and faster and faster until it was a blur. He apparently smelled or heard the hunters returning, for in less than a minute over the ridge and into camp they came. With that, the omega, relieved at last of his rambunctious charges, collapsed to the ground exhausted.

During their fourth week of age, the pups begin feeding

on partially digested food disgorged by the adults. The stimulus for the regurgitation can be the mere presence of pups or their begging actions—licking the corners of the adult's mouth, nuzzling its face, nipping at its jaws, or pawing the mouth and head. Age and parenthood have nothing to do with the response of a wolf to a pup's begging. A pair of year-old tame wolves disgorged food immediately upon seeing a litter of pups.

The babyish behavior of pups is the source of adult lick greeting, but you can notice with dogs that puppies are much more insistent on licking your face than they will be when mature. They try it on everyone, not just those few people to whom they will later be devoted.

## WILD AND DOMESTIC SIMILARITIES

Children in a human family are tended by dogs much as wolves tend their young. They set themselves up as baby-sitters, and, if permitted, most will accompany youngsters to the school bus and out to play and will sleep on their beds to guard them like the mother wolf does. They also take all manner of abuse from these human pups, as does the wolf baby-sitter. I have seen little children pull the family dog's hair, bite it, pummel it, and roll on top of it without a word of protest from the dog, just a wry look of "This is how it is." Babies inspire special care from the family dog. My bluetick hound, Gunner, on hearing one of my infants cry would rush to me and stare until I did something about it. She guarded the baby buggy when an infant was in it, and although she never regurgitated food, she did heave the first day I introduced her to my screaming infant daughter.

Sometimes a wolf or dog will threaten unfairly and then apologize. In the Barrow lab I saw a pup run up to its mother and crawl between her legs to nurse. She turned and snapped but in the confusion of puppies disciplined the wrong one. She promptly apologized by smiling—pulling her lips back and opening her mouth slightly—and wagging her tail once. Dogs also apologize. When my Airedale snapped at me for trying to doctor a wound, she apologized for her aggressiveness by quickly relaxing her ears and opening her mouth. She wagged her stump tail—once. I accepted her apology by smiling, too, and shaking her muzzle in friendly dominance.

Two golden Labs that belong to my friend Kathy Kunhardt came into the living room where we were talking. Their ears were pressed back, their foreheads innocently smooth, and their tails wagged now and then, but very low.

"What are you apologizing for?" Kathy asked them and hurried to the kitchen. "They've taken a whole wheel of Brie off the table and eaten it!" she called in anger.

Kathy did not accept their apology but disciplined them with snarl words: "No, no, bad dogs. Bad." She stood above them to say this and shook her finger. She frowned and stared hard at them. They dropped into the submissive standing pose of the omega, pulled their tails in between their rear legs, lowered their heads, and drew their ears down, subdued—at least for the moment.

Discipline is the major subject in a wolf pup's schooling. It must learn what it can and cannot do, not only for its social welfare within the pack but for its safety on the trail, which is fraught with dangers: other wolves, human hunters, goring prey, bone-breaking crevasses, freezing weather, and long periods of starvation. Lessons begin early. If a pup wanders off, it is brought back by mouth. If it persists in straying as it grows older, it is growled at, then snapped at with teeth showing. Never does a mother wolf or dog bite a puppy. The threat of biting, conveyed by the bared-teeth threat face and the snapping gesture, subdues the pup as did Kathy's harshly spoken words and shaking finger. A pup is also disciplined by the mother standing above it, as she will do when siblings are fighting too strenuously. Her mere towering presence is sufficient to break up serious battles for dominance. A paw placed squarely on a pup's shoulder also subdues it. Once the rules are learned, the mother need only use her eyes, like the alpha male wolf, to get obedience.

Mother dogs use exactly the same vocabulary to discipline their young, and we use variations that share many of the aspects of canid talk. Kathy growled out her words, and she arose and stood above her dogs. Her lips were pulled down, and her teeth showed on the word "no." More important, she accepted the dogs' apologies sometime later, and the friendly bond characteristic of pack and human family was reestablished.

Both the discipline that you may have to give your dog and your acceptance of its apology will proceed with decorum if you remember that you are its alpha.

### ON BEING AN ALPHA

And you *are* the alpha. When you take a pup from its parents or adopt an adult, the dog turns its love of the leader upon you. It needs no reward for this gift other than praise and affection. It has one happiness: to please you. You do not have to strike your dog for doing wrong. Your frowns, finger shakes, and sharp words are enough. You do not have to reward your dog with food. Your praise is its laurel. It wants to do what you wish it to do, to follow your commands, to hunt for you, guard your property, lie at your feet, be near you. The more you behave like an alpha, the more smoothly the relationship will go.

"How does a wolf alpha come to be?" I asked Dr. L. David Mech, the U.S. Fish and Wildlife Service wolf authority who is at Ely, Minnesota, studying some of the last of the great gray hunters in the lower forty-eight states.

"Very naturally," he answered. "They become alphas by becoming parents. Wolf pups, like children, look up to and obey their parents." That is the general rule, but in well-established packs many wolves do not breed, or if they are permitted to do so by the alpha, they are still subordinate to him, although boss to their own pups. Only when an alpha dies or becomes too decrepit to hold his position does a wolf get a chance to become the leader of the pack. To hold the position the alpha must constantly work at it. In an experiment an alpha removed from his semiwild pack for two weeks and returned at the end of that time never regained his leadership.

Some leaders reign for twelve, even fourteen, years, and during their tenure they may dominate the other males in their pack to prevent them from breeding. Just who these other members are was not known until Mech put radio collars on the wolves of Ely in the 1970s. Most are family, grown pups who have not left home, but some are young individuals who have taken off from their own families and have been invited to join another. Whether they are kin or foster members of the pack, they have to replace the boss if they are to become parents (and therefore alphas). The other alternative is for a male and female to go off and start their own family, in which case they are automatically alphas, but without the hunting support of others, that is difficult to do.

Dr. Michael Fox, an expert on wolf leadership, explained to me that puppies sort themselves into a hierarchy even before they struggle for status among the older members of the pack.

"There is an alpha in every litter of puppies," he said. "They emerge during the second, third, and fourth months, but sometimes as early as ten days, as the pups fight for dominance among themselves."

The alpha is always the largest, but when sizes look the same to you, the leader can be found by putting a bone in the midst of the pups. Dogs never fight over milk or any liquid and seldom attempt to defend a dish in which dry food is always present, but they will fight over a real bone from two weeks of age until the question of who is boss is decided at about eleven weeks. At that time the pup that has most often won, usually a male, becomes the alpha. After that time there may be no fight. The alpha will get to the bone first, and the others won't argue. The bone itself confers safety from the others' aggression. If you hold all the pups back but the omega and let it get the bone, the others will keep their distance even when you release them. It is understood, among dogs as among wolves, that food belongs to the one that claims it first, no matter what its status.

"All fighting stops when the alpha is established," Dr. Fox continued. "You can recognize him. He is fearless, initiates all the activities, and sticks to a job longer than the others." As a schoolteacher commented when I recounted these facts to her, "There's one in every classroom." She wasn't talking about bullies. A sign of the alpha is that he doesn't have to fight to get his way, and you, as an alpha, don't have to fight with your dog to get it to obey. The obligation of an alpha is to teach.

Dog training is not cruel and does not suppress individuality as some sentimentalists believe. It is an extension of the training a canid receives from its parents. Canid parents discipline their offspring with voice commands, gestures, facial expressions, eye contact, and combinations such as the threat stare, which is akin to the look your mother gave you when you were caught with your hand in the cookie jar. And they discipline for the same reasons we do—to make their offspring good citizens that are happily adjusted. There is little democracy among the social canids. They are alpha oriented, and happiness is doing what the leader wants them to do.

A well-trained dog is more alert, more communicative, and happier than one that is permitted to become a pest that must always be reprimanded for jumping on people, running away, eating off the table, knocking things over, and stealing objects from tables and counters. The dog that knows its alpha wants it to move back from the coffee table at cocktail time and lie down when told to do so is in tune with its heritage. You are the benevolent alpha, and all stress and argument ceases.

But people feel more conflict about pulling rank on a pup in order to teach it the rules of its family pack than an alpha wolf does. Edward Fouser, one of the first Seeing Eye dog trainers, who now specializes in training master and dog to communicate and work together, can train a three-month-old pup to heel, sit, and stay in one hour. It takes him eight hours of lessons when working with owners.

"I spend more time training the owner than I do the dog," he said. "Dogs learn quickly when they know what you want and are firm about it."

Whenever possible Fouser uses dogs' own language on them. He teaches a dog to stay within the confines of an area by placing a rope on the ground across driveways and escape routes. He leads the dog to the rope, pretends to trip over it, and yelps like a dog in pain. That yelp does it. The dog draws back and drops to its belly. Fouser picks up the rope, shakes it in front of the dog's face, and repeats, "No, no. Bad. Bad."

The dog does not move. Fouser steps across the rope and waits to see if the dog will follow. Most stay where they are, not crossing the forbidding and frightening barrier. If the dog does cross, Fouser snaps the collar and says, "No," again. That usually does it. Even my malamute, Qimmiq, who hates to stay, did drop to his belly and remain behind that rope, and to this day he remains behind other ropes he sees.

## SURVIVAL SENSE

At least a dozen dogs in my hometown have been trained by Fouser to stay behind their boundary ropes even when they are covered with snow. There may be an additional motivation besides the yelp.

Wolves avoid murderous encounters with strangers from neighboring packs by leaving between territories a buffer zone where no wolf trespasses. L. David Mech, studying his radio-collared wolves at Ely, also discovered that the wolves do not hunt in these buffer zones. Because they don't, the buffer zones have come to serve as game preserves. They keep a few deer alive in times of stress when deer populations drop low. After killing off the animals on their territory, the wolves starve and die rather than hunt the buffer zones. The deer, protected from the wolves by their own social rules, begin to recover.

"The buffer zones," writes Mech, "tend to . . . form a reservoir for maintaining and recovering deer populations."

Keeping enemies at a distance with a no-man's-land keeps the predator-prey relationship balanced. And so Ed Fouser is able to teach dogs not to cross a rope—no-trespassing zones are not alien to their kind. You can also teach them to stay out of certain rooms by calling on their respect for boundaries.

The yelp is like the alarm cry of a bird that sends the flock to safety. I saw the mother wolf of Sanctuary River come romping with her pups to the top of a ridge within view of Gordon Haber and me. She looked down, saw us, and let out a wild high-pitched yelp as if in pain. The pups instantly turned and ran. A few moments later three pups returned to take another look at the curious two-legged animals. The next yelp from the female was bloodcurdling. After that we saw no more of the pups.

The yelp can be overused on dogs. They are canny, soon catch on that your "alarm" is insincere, and make their own judgments about a situation. But used for situations of genuine alarm it can be very effective.

**THE ABCs**

Dogs are clever at detecting insincerity, and if you are caught in duplicity, you are likely to endanger your status as an alpha. For instance, don't laugh at the mistakes your dog makes during training. It knows what a laugh is all about—fun—and with that all discipline breaks down. Qimmiq went spinning in circles when I laughed at his defiance of my request that he sit.

"Don't laugh!" Fouser admonished me. "The dog thinks you don't mean what you are saying—that lessons are funny. They're not." I snapped his collar, firmed my attitude, and Qimmiq sat.

Just as wolves send messages through eye contact, so do dogs, and so can you. When you are pleased, look into your dog's eyes. It will see your pleasure. But be sure of your mood. If you are angry with it, you will be giving an intimidating threat stare even if you are saying, "Nice doggie."

A kind gaze also helps a pup learn its name. Say the name while gazing into its eyes, brightly and with a smile. Repetition reinforces the identity you are transmitting to the pup through voice, eye contact, and your smiling, attentive face. The pup puts the whole thing together and arrives at a sense of self as early as two months. Now when its name is spoken, it looks at you. Eventually, if you are just talking about your dog in ordinary conversation to another person, it will prick its ears and listen. Dogs like to be talked about.

So easily does a dog learn its name that one wonders if wolves do not have names for each other, special barks to attract the attention of a specific individual. Certainly they have names for themselves. A wolf calling from a distance is immediately recognized by the members of its pack as well as by the members of neighboring packs. Who the wolf is determines how it is answered: ignored, howled to, or sought out.

The name you give your dog is important and should be picked carefully. Dogs like the sounds of hard "g" and "k," or at least they respond more quickly to these sounds than softer ones when tested. But the name should be to your liking and one to which you can give a variety of inflections, from soft to firm. "Butch," for example, is difficult to soften; "Willa" is hard to say firmly.

Dog trainers who have become famous for the apparent ease with which they get dogs to listen to them are like the great black alpha I watched that summer on the tundra. They are tolerant and affectionate, but more than anything they are exciting. Writes Barbara Woodhouse, the famous British dog trainer who dares to train dogs before the TV camera: "Giving praise and affection is where a multitude of owners fail their dogs. A pat and a kind word are not enough in the initial training of dogs; the atmosphere must be charged with a certain excitement, for dogs are very sensitive to excitement; when they have done right they love the wildest show of affection and a good romp. Dull owners make dull dogs."

Training is expedited by surprise. Trainers use the choke collar not to hurt but to snap pup or dog to attention. Keeping their attention is half the battle. Dogs tend to daydream, and the collar snap brings them back to reality. Said Fouser, "Training is communicating." You can't communicate with a daydreaming dog.

But you also have to know when not to invite a romp. Don't invite your dog to a game of chase when you are trying to teach it to come. To chase is to say to the dog, "Run away." Dogs chase when they see something fleeing, and they flee when something chases. Call "Come" and stand still. This takes enormous self-control when you see your dog ignoring you and walking off, but it works. Be the alpha, call and stand still; that is dog talk. Fouser can teach this command in a few lessons.

Dogs learn our commands so readily when they are taught correctly that they must have their counterparts in dogese. The first time I told the pup Qimmiq to stay, he did so—and I was mystified until I saw a golden Lab mother stop a pup from pestering her. Her face, gestures, and attitude were not unlike mine—mouth slightly open, eyes wide, feet firmly

planted, and body stock-still like a wolf that, detecting possible danger, freezes. I also held up my hand, a signal that dogs come to know even without a verbal command.

To teach your dog "Come," "Heel," "Stay," "Down," and "Off"—the basic commands every dog should know to be a good citizen—is not asking any more of it than the alpha asks the wolves of his crew. A look or a tail wag will summon a pup to its mother. The alpha glances over his shoulder at a juvenile that is gaining on him to hold it at "heel," and the "stay" signal can be a threat stare or a growl. To tell a lesser wolf "down," the alpha need only arise, lift his head and tail, and stand tall. Snarls suffice to keep a pup "off" the tail of a resting mother wolf. With these wolf words in mind I watched Fouser train a dog to heel. With a jerk on the choke collar, he got her attention. He gave the command and stepped forward. The dog pulled back. Fouser looked at the dog, jerked the collar, and looked into her now alert eyes. "Heel." His voice was quiet and low. He encouraged the little three-month-old when she came along. He was absolutely gentle, but firm. When she began to track along beside him, he used his wiggling fingers to keep her attention. Seeing the movement, she was diverted from a sudden preoccupation with another dog, looked up, got the friendly eye, and trotted on. Several months later I saw the dog heeling as her mistress walked along the road—no leash or choke collar on. There was a happy dog, I thought, following a strong leader because she had been told what to do and was pleased to find her role.

## IN LOCO PARENTIS

Pups are not born knowing everything about how to be a dog, much less how to be man's best friend. They have much to learn from their parents. Although an old dog *can* learn new tricks, there are certain critical times when dogs learn certain things best, or most easily. Since we remove puppies from their mother (and they may never see their father at all) by as early as the sixth week, we are, like it or not, in loco parentis for a pup's first year.

Dog communications begin at birth when the mother, by licking her pup, stimulates it to breathe. First-time mothers may fail to give this crucial message; you may have to "tell" the pup to breathe by gently rubbing it with a terry-cloth towel.

The scent of the nipple is probably the second communication, a directional signal that guides the pup to nourishment. If unable to get the nipple, the puppy may utter its own first word—a high-pitched whimper, a call for help.

The signal for helplessness originates in clean-up time. The mother stimulates the pup to eliminate by licking its anal region, cleaning the puppy while nourishing herself. The pup usually rolls onto its back during this ritual, a pose that later

in life is a special mark of the omega but is also a dog's white-flag signal, which says that it has given up in a fight and, like a helpless baby, is not to be hurt.

The first few days of a puppy's life are spent feeding and sleeping. The mother does not leave the young for at least a few days, except to eat and eliminate—somewhat reminiscent of the mother wolf.

The eyes open between seven and ten days of age, and the pup wobbles forward on belly and feet. Probably by relying on its sense of smell, it can identify mother and keeper and can avoid strange objects. It yelps to say it is surprised, hurt, cold, or hungry. The mother usually responds to these cries, but people can begin talking to a pup by answering the needs. Pick it up and reassure it if something has surprised it. If it is cold or hungry, snuggle it back among its siblings and call the mother.

Dogs are not born loving us. This is taught by their mothers and handlers between the ages of four and fourteen weeks, the critical period when a dog becomes happily adjusted to people or remains forever shy and antisocial, like a wild thing. Pups raised without human contact flee with their tails low at the sight of people and are as elusive as wolves.

On the other hand, a pup that is taken from its mother and fed on a bottle before its eyes open becomes what breeders call "humanized." The pup ignores other dogs and behaves more like a person than a dog. When Stew, a bluetick hound, was six days old, he was put on a bottle by Emmy, the wife of George Norris, a fox hunter of Seneca, Maryland. I met Stew when he was about five years old. He did not bark at or run out to investigate other dogs, but sat on the porch and directed his attention to people's conversations, turning his gaze from face to face as the discourse flowed. He got to his feet when Emmy got to her feet, and he drank from the kitchen sink—no tin bowl for him. He slept in a bed and wagged his tail when Norris laughed, as if he shared the fun.

Willa, a mixed breed who belongs to my friend Melissa Young, is another bottle-raised dog who also thinks she is a person. When Melissa's friends come to visit, Willa walks serenely out of the house and greets them with a face smile—lips pulled back and up to expose all her teeth as people do when they smile. She was not taught this; she learned it herself. She wrinkles her nose deeply, much as Melissa does when greeting her friends, and she escorts the guests to the living room. She sits down after they sit down.

At night Willa takes a "blanket," or comfort rag, to bed with her, preferably one of Melissa's old shirts or socks, which she sucks until she falls asleep. When Melissa's brother Andy visited her in North Carolina one spring, Willa took a shine to him and followed him everywhere, smiling, wrinkling her nose, and wagging her tail. Then Andy departed. She became depressed, her head drooped, her tail was down. Just before nightfall she passed the clothes hamper in the hallway, stopped, tipped it over, and, from the jumble of shirts, socks, and blue jeans, pulled out a T-shirt Andy had forgotten. She carried it to her bed, sucked on it, and was comforted.

"And she's very easy to talk to," said Melissa. "I simply explain what I am going to do with words and gestures—go to school for instance—and she trots to her bed and lies down.

"Occasionally she argues with me when I tell her she can't go to the store or that she's been bad—chewed up a shoe or something—but for the most part she understands and agrees."

Melissa did not pick this pup deliberately—Willa's mother had been killed by a car, and she took in the orphan. Breeders and behaviorists agree that the best family dogs are those raised by their mother with their siblings in a home where kids and adults pick them up and play with them. These pups develop affection for people and are interested in and relate to other dogs, being neither wild nor overhumanized.

During a puppy's early weeks its smelling apparatus is developing rapidly. It will approach a person slowly, nosing and sniffing shoes and clothing. A hand offered for smelling is an appropriate response, because scent "conversations" help to bond the puppy to you. The pup recognizes the individual odors of its siblings and parents as well as those of the humans who handle it, and it may cry if picked up by a stranger before it has had a chance to "sniff him out."

At this point the mother less often takes care of her puppies' elimination, and they will leave the nest to wander with their noses to the ground as they search for old odors of urine and feces. When they find such marks, the scent triggers the pups to eliminate. They don't care that the mark is on the rug or in the middle of the living room floor. The search can be a communication and can be used to housebreak a pup. As soon as it explores with its nose to the

ground, tell it where to eliminate by picking it up and putting it down on paper or outside. Paper with a trace of urine works wonders, the scent acting like quarters in a coffee vending machine.

One tainted paper can last a long time. I brought Qimmiq home a year after my Airedale, Jill, had been put away. The very first morning, he smelled an old mistake Jill had made on the rug as a pup twelve years before, squatted, and peed. There is only one remedy for that long-lived, urine-scented communication. Replace the rug.

About ten days after the pups leave the nest to eliminate, *following* behavior begins. The pup, without being given a command, instinctively follows its mother or another pup or a person for short distances—reminiscent of the wolves following behind the leader on the tundra. Speak to this development by leading the pup without a leash and repeating the word "heel." The earlier it hears this the better, even it if is not connecting the word with what it is doing. Your enthusiasm for its presence at your side can extend the behavior for longer and longer distances.

Although a pup learns from its mother from birth, beginning at about four weeks it has new capabilities for learning. By her own reactions the mother teaches the pup discrimination. It will turn its head at the sound of a familiar person and run from a stranger. It is now capable of making judgments and can discern the difference between danger and those things without significance. Evidence that enemies and dangers are taught by the mother comes from pups separated too early to learn such lessons. Willa, the bottle-raised pup, did not respond to a fire until she got a paw singed, whereas Qimmiq, who had lived with his mother for three months, drew back from fires but not from a hot electric plate, which his Arctic mother didn't know about. He was afraid of engines starting up but did not notice banging doors. Television was a harmless curiosity, which he gazed at without concern even when it crackled with the sound of gunfire. His mother had been trained not to shy at such sounds. This period of intense learning, which lasts about a month, is a good time to intensify communication with a pup. The month-old pup can bark to demand attention or announce a stranger. Answer its request for attention by stroking it with the same soothing movements of its mother's tongue. Praise it for announcing a stranger with a pat and some enthusiastic words. It can use its sense of smell to track down food and family—your clue that it is time to call the pup to dinner or to play. The pup's feelings become easier to read. It can wag its tail to acknowledge living things. It can wrinkle its forehead, one of the first of the complex facial expressions that will develop.

## THE MAKING OF A BEST FRIEND

Between the fourth and thirteenth weeks is the best time to establish a relationship with the pup, or, as behaviorists say, *socialize* it. Pick up and fondle the pup every day, talking to it to make it comfortable with humans. Early handling will save months of work later on, and some pups cannot be socialized at all if the process is begun after four months, when it may be too late to interact with people.

Aggression can be handled right out of a pup's personality in this same period if a mild-mannered dog is what you want. This is the time when siblings are fighting for dominance. Behaviorist John Paul Scott and his students at the University of Chicago stopped pups from fighting by picking them up every time they tangled. Diverted by being off the ground, they instantly quieted down. After the tenth week they did not fight at all. By six months they were unusually nonaggressive pups. "Don't fight," you are saying by picking up a scrapping pup during this period.

On the other hand, letting them fight will give you a clearer idea of which is going to be the alpha, which the omega. The alpha will eventually fare best in the family that appreciates its alertness, courage, initiative, and stick-to-itiveness. Both male and female alphas learn quicker and are more intelligent. An omega makes a better dog for working people who do not have much time to interact with it. Omegas, which are often the smallest in their litter, are pleasant dogs, relatively nonaggressive, and often shy. Omegas make lovable pets.

Many people consider six weeks a pretty good time to find homes for puppies—perhaps because the mother is neither nursing them nor cleaning up after them at that point, and the alpha human keeper is degraded to the role of a babysitter. But five to six weeks is a particularly vulnerable time for the puppies. They are learning to fear those things their mother fears, as well as developing a few fears of their own. They will draw back or run away when a person walks rapidly and noisily toward them, even though they had not reacted in fear to that person a week before. Each pup is also becoming an individual, and its new ability to make independent judgments is equaled by its new susceptibility to psychological damage based on its unique experiences. If it is hurt or badly scared by people during these weeks, it may never overcome its fear of people. With this vulnerability comes a strong attachment to place, so if it is taken from its home, it is all the more unsure of itself. Your word to the pup now is gentle hands and slow, not fast, movements.

If you are going to keep a puppy yourself, use the strengthening of its interest in following through weeks five to nine to introduce the leash. The pup will track along nicely

## PUPPIES' SOCIAL PROGRESS CHART

*During the first 4 weeks the puppy is more interested in its mother than in people.*

| | |
|---|---|
| Week 1 | Sleeps, nurses, yelps when hungry, hurt, or cold |
| Week 2 | Eyes and ears open, lifts belly from ground to walk |
| Week 3 | Leaves nest to urinate |
| | Yelps when restrained or when removed from mother and litter mates |
| | Gives earsplitting yelps when removed to a strange place |
| | Fights playfully with litter mates |
| Week 4 | Follows mother, litter mates, or person |
| | Approaches a person slowly and wags tail |
| | Startles at sudden sounds and movements |

*Weeks 5 thru 13 are the period of socialization when puppies that are handled daily adjust well to people. The most critical period is week 5 and week 6. Ups and downs characterize this time, so leash training is best done during alternate weeks 5, 7, and 9. House training can begin during week 8.*

| | |
|---|---|
| Week 5 | Weaning begins |

without being yanked, but do encourage it to walk on your left, and use the word "heel" with spirit and enjoyment.

Pack movement begins around two months. This is not just following. Suddenly, without being led, all the pups rush to the door or gate in one tumbling mass. Since we discourage pack movement in our dogs, there is no reason to talk to this development. Observe and enjoy it; a mass of puppies coming toward you is wonderful.

Two months, or even a bit earlier, is the natural time to teach the word "no." During weaning between five and

| | |
|---|---|
| | Learns first fears from mother |
| | May become fearful of handler |
| Week 6 | Becomes curious about new people and dogs |
| | Becomes emotionally sensitive and may be susceptible to psychological damage |
| Week 7 | Weaning is completed |
| | Begins to attack other pups by ganging up |
| Week 8 | Uses definite spot to defecate far from food, will not defecate if shut up overnight, and sniffs for a spot to urinate where there is a trace of urine scent |
| Week 9 | Chases and catches other pups and toys |
| | Fear of handler disappears with daily handling |
| Week 10 | Hierarchy within litter established, serious fighting ceases |
| Week 11 | Hunting behavior begins |
| | Anxiety in strange places eases |
| Week 12 | Makes the best transition to a new home |

*For better or worse, 13 weeks marks the end of the period during which a pup can be optimally socialized. By 14 weeks, when the pup is accustomed to its new home, training to "Heel," "Sit," "Down," "Stay," and "Come" can begin.*

---

seven weeks, the pup has learned the meaning of "no" from its mother. Hers is the growl-bark plus the wrinkled nose and the drawn-up lips that expose her canine teeth. She snaps close to a pup's face but never bites. You can come pretty close to the dog "no" by lifting your own lips in a menacing grimace and jutting your face toward the puppy as you growl-bark, "NO."

At twelve weeks hunting behavior is emerging, and with it a readiness for new experiences. That's the age when the wolf parents move their offspring from the nursery den to summer camp, when the puppies begin to exuberantly practice hunting games on the patient baby-sitter, and when they begin to participate in the greeting ceremony that will bind them to their alpha and the pack.

Breeders agree that sixteen weeks of age is the ideal time for puppies to leave for their new homes, not only because the puppy is more sure of itself by then, but because prospective owners can tell more about its personality.

Dog pups, like wolves at the same age, lose at this point their strong attachment for the first home place—enthusiastically practice hunting skills with balls and sticks—and look to whoever seems to be in charge for love and leadership. This is not necessarily the person who feeds the pup, as many people think, but the one who most clearly teaches the puppy how to behave, and that can be a child.

Not that leaving home and pack is easy. When I took Qimmiq from his mother, he buried his head in my arms and kept it there as I walked home through Barrow's snowy streets. He would not look at me. He was eight weeks old, a month younger than the ideal time for separation, and stoic. He made no sound when I put him down on the floor of the house. After a long look around him he searched for his mother, nose held high as he ran sniffing for her scent. I petted him and called his name, but he acted as if I were not there. His eyes had the look of the daydreamer, seeing without seeing.

The next day he gave up the search for his mother but still buried his head in my armpit when I picked him up. I talked to him, whimpered to him, sang to him, and after a few minutes he slowly lifted his head and focused one yellowish brown eye on my furless face and then on my blue eyes. He seemed to be registering me on his consciousness at last—and I was in his world. He cuddled against my body.

Thereafter he was a domestic dog.

## THE EYES HAVE IT

I have become convinced that eye contact with dogs, as with people and wolves, is loaded with meaning. Qimmiq would not give me his gaze until he was ready to shift his love to a new leader figure, but once he took me in through his eyes and bestowed upon me those deep instinctive feelings that are love, his flashing glances kept us in contact from then on.

When I take Qimmiq to the woods to let him run and bound over logs and rocks like a deer, he comes back, gets my eye, sees me nod, "I'm okay, go on" while looking him in the eye, and then turns and dashes off again. When strangers enter the house, he looks at me, I signal with my eyes, "This is a friend," and he comes forward wagging his tail. Whenever I go in and out of the front porch where he spends much of his time, I take his chin in my hand and look lovingly into his eyes. He looks back, his ears go down in deference, and his tongue comes out as if to lick. He loves to wander like his near kin the wolves, but I can get him to come back to me if, when I call his name, he turns toward me. I catch his eye and lock his gaze with my threat stare. With reluctance, for he is still a pup, he slinks toward me. If he really does not want to come, he will avoid my eyes and bounce on as if he did not hear.

I tell him many things with my eyes: that I am busy and cannot play, that he cannot come with me in the car, that he is beautiful, that I like to have him with me. You don't need words, although I often use them, for, to a dog, your eyes brim with meaning.

## THE LANGUAGE OF SCENTS

Many puppies announce their adolescence at about six months by taking a first long excursion from home. This urge can be indulged where there are no leash laws, for the explorer will come home. However, Coco, a brown poodle in my hometown, off on her first juvenile excursion, was picked up by the dogcatcher, and her mistress was fined twenty-five dollars.

Adolescence lasts for two or three months, culminating for the female in the first estrous period at around eight months. Estrus, the period that ends in several days of fertility, is marked by the emission of blood and a change in behavior. The female is more affectionate, whimpers restlessly, and urinates more frequently than usual.

The male adolescent pup begins lifting his leg to urinate in the manner of the adult canid. Although he may long ago have settled dominance matters with his peers, his elders no longer consider him a "mere puppy," and he may have to scrap with other dogs in the neighborhood until his position is established in this wider world.

Both the sexual status of females and the social status of males are advertised by scent. A dog's smell receptor cells bear a number of fine hairs at their surface that pick up chemicals and report them to the brain. A dog has more than a million of these hairs, giving it the capacity to discriminate among thousands of smells in doses many hundreds of times more diluted than a human could detect. On a scale of one to ten, our ability to pick up odors is two compared with the dog's ten.

Scientists believe that scents tell dogs more about the world than any of their other sensory organs. We have such poor noses that we are ignorant of almost all of the information dogs read as easily as we read a newspaper. Chemicals that lie in footprints tell the species, sex, and emotional state of the printer; the odor of a storm blowing on the wind sends the reader to shelter. Unfortunately, that avenue of communication is a one-way street to us, for although we apparently communicate information about ourselves to dogs, scent is the medium we understand the least. We stumble along on eyesight, while our hound presses its nose to the ground and runs off through the woods on the scent of a deer as if it were following Interstate 80.

I learned about the odors of the field and forest from Gunner, my bluetick hound. They are exciting, she said with her hysterical barks. They are also hot, cold, in the air, on the ground, old and faint, new and vivid. As I watched her run and listened to her report, I thought that to her those invisible scents must have been presences with shapes and textures. She said so with her voice. A hot trail was a high yipe, yipe; a cold trail was a bark given now and then. Once, she announced with a series of yipes a scent so fresh and inviting that I dropped to all fours and sniffed the earth. There was no odor of the furry footstep of a rabbit for me, let alone the aroma of its being startled from sleep or the smell of its fear as it fled—only the scent of dry earth.

Realizing that scent is a foreign language to us, we have bred dogs to be our translators when we are in the field and forest. Their voices ring out announcing the presence of game and direct the human hunter to it. This communication between hound and human did not come from the wolves—they hunt silently. Wolves approach their prey without a sound to surprise it. As they move, each one keeps its eye on the leader and, in silence, makes up its mind what part it shall play in the stalk and kill. Even the kill is quiet. A good voice on the trail was selectively bred for in the dog. A call from the briars and dense thickets, "I'm on the trail of a rabbit," was much more valuable to us than wolves' noiseless strategies.

Dogs do translate by bark the kind of prey they smell, and some also listen for our spoken commands to know what prey we seek. Having a dog is the next best thing to having a good nose. My father used to discuss what was afoot in the Potomac River bottomlands with his hound dog, Spike. He would tell him, "Hunt up a squirrel," and Spike would circle the forest until he picked up a squirrel scent. He would run the animal up a tree and announce it until Dad caught up. Spike found a raccoon when my father wanted a raccoon and deer when he wanted a deer. When they were simply out to investigate the forest and learn what was around, Spike would tell Dad what scent he had come upon. He had a different voice for each animal. A rabbit was a high yipe, a raccoon a series of sharp barks, a squirrel a low yip, and a deer—well, that was a rapidly fired succession of barks, yipes, and bellows, a dog's name for the quintessential odor.

### PUBLISHING THE NEWS

That is the receiving side of scent. On the transmission side is *scent marking*—the leg lifting of males, the urine ads of females in heat, the depositing of feces, and the spreading of gland secretions that broadcast to other dogs just who the marker is and what it is up to.

Scent marking is the paramount language of the canids. Over the eons it has become ritualized, the male lifting his leg to shoot his message high and far, the female, or bitch, squatting. One of the canids, the bush dog, carries the ceremony to the extreme. The male walks his hind feet up a tree as high as he can while balancing on his front feet, then squirts.

The ritual of the domestic dog includes methodically marking vertical objects, such as trees and fire hydrants; odors on the ground, such as that of a carcass or the urine of a female; and new things in the environment, such as a paper cup that has been recently dropped, a newspaper pitched up on the porch, or a new chair set out in the yard.

The male uses three positions: standing with one hind leg raised and crooked when squirting on a vertical object, standing with all legs spread apart to mark the ground, and the same pose with one leg slightly lifted. Females, who scent mark less frequently, sometimes lift a hind leg under the body as they squat.

Since we can't really know what dogs are saying in these messages, animal behaviorists have come to conclusions by observing what they mark and how other dogs react to the message. Said Swiss ethologist Rudolf Schenkel, one of the first to study scent marking, "Canid scent marking is a social

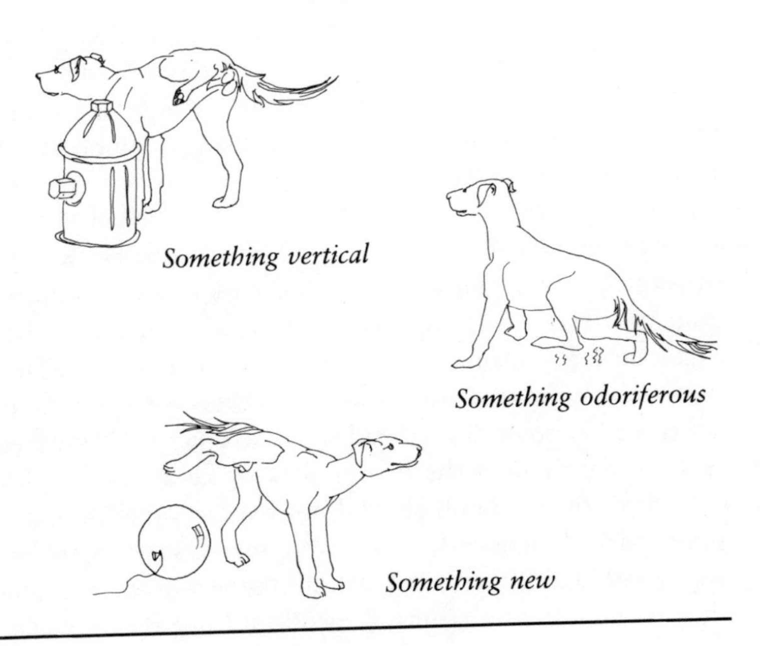

*Something vertical*

*Something odoriferous*

*Something new*

act that defines the territory, makes acquaintances, forms pairs and legitimizes the alpha's position."

British scientist M. Lyall-Watson concludes that an animal's scent on posts and objects maintains its familiarity with the countryside. A dog that roams a street each day seeks out and finds its own scent on a garbage can or automobile tire and is reassured by the familiarity of the posts it has "colored." It jogs along loosely, confident. If, however, it steps into new country, it becomes nervous and marks liberally to make the unknown familiar. Its making odoriferous marks is like our blazing the trees in a strange forest to find our way in and out. The scent reassures the dog as the blaze reassures the man, and the landscape becomes as nice as the sight of your street after a long trip.

Scientists think that the canid family can tell from urine scent alone who the urinator is, its sex, age, health, what it has been eating, and whether its mood is angry, fearful, or content. After sniffing up a news report from a neighboring dog, a male—less often a bitch—tops that with his mark and information of his own. The fire hydrants and lampposts of the block are gossip columns for dogs.

And, like wolves, dogs mark a territory. The borders are learned from you. When you step off your property, the dog senses a change in your attitude. It will bark to warn strangers from your home turf but will not bark as ferociously, if at all, on the sidewalk in front of your house or apartment. That is common ground. When I walked Jill off the property onto the road, her smooth forehead tightened, her neck lowered ever so slightly, and she became less frisky as she lost her confidence off her land. She did not bark at other dogs and scent marked frequently. On her property she urinated to empty her bladder but rarely let fall the few drops at a time that are typical of scent marking. When I put her in the car, however, her self-assurance returned. I had conveyed somehow through my attitude that this was home turf, too. She barked as murderously at dogs and people from "her" car as she did from "her" house.

Evolutionists speculate that scent marking grew from a fear reaction. Young puppies removed from their nest to a strange place urinate out of fear. It seems likely that that is a remnant of the behavior primitive canids practiced even as adults. Eventually the urine mark became useful. Other canids coming upon it either kept their distance or, looking for a mate or companion, approached. As for the marker, when it circled back and found the smell of itself, it was reassured. The foreign territory was familiar. It had been there before.

As scent marking evolved, the male canid ritualized the behavior by lifting his leg, probably to hit high and show how big he was, and by saving squirts to top another's squirt, a crude measure of status. Today, an individual's blatant newscasts on a city fire hydrant can read, "I am big. I am male. I am young. I am confident. I am irritable. I am off my property."

## TALKING SCENTS

Given the power of scent communication, there is scarcely a naturalist who has not tried to speak to wild canids through this medium in and around the campsite. Farley Mowat, author and naturalist, is one of the more successful experimenters. Observing that wolves scent marked the perimeters of their property to keep off other wolves, Mowat marked off an area around his Canadian campsite in the manner of the wolf. As he reports in his wonderful book *Never Cry Wolf*, the wolves got the message and respected his space. L. David Mech, on the other hand, who was also inspired to scent mark his campsite while studying the wolves of Isle Royale, Michigan, got nowhere. "They paid little if any attention to my mark," he said.

Dogs also ignore human urine. No one knows why. Maybe humans say nothing in their urine, or maybe dogs know we can't smell their answers anyway, so why bother?

The number of times a male scent marks, most dog owners agree, is an indication of just how macho the pee-er is. A terrier named Butch was taken to Japan by my neighbors Sis and Jay Melvin. When Sis walked him in the streets of Tokyo, he marked objects constantly, "Every foot or so," she said. "It took forever to walk a block. One day I kept tabs. I counted 109 pees, and he was still going when I finally dragged him home. What a male." But I suspect his many pees also indicated his nervousness on those foreign streets, for dogs scent mark frequently when under stress.

Feces marking by dogs is both odoriferous and visual and, consequently, a louder statement than the liquid one. This mark is not used as frequently as urination, feces usually being hidden. When it is used, however, you cannot miss it, for it is put down to attract attention. Dogs competing for our female left their visual marks on stones, in the middle of the drive, and even perched on flower pots.

Scientists consider feces display to be a part of scent marking that also arose from automatic responses and evolved into a way of familiarizing the animal with the environment. That, they agree, is the first function. Secondarily, feces marking, like urination, is a social display that brings the sexes together, maintains the territory, and, therefore, is an important function in the survival of the species. That, I agree, is why my yard was so odorously decorated.

Anyone who has walked a dog on a leash and tried to curb it when other dog feces adorn the sidewalk has seen the visual importance of feces. The dog sees, sniffs, identifies the depositor, and carefully selects a spot nearby to put its mark. Both males and females scratch the ground with forelegs and hind legs after feces marking, but it is more common in males. With firm strokes of their feet they leave these inscriptions for their neighbors to see. On loam and in grassy yards the scratches and feces become a signature. I can identify the marks of two dogs in my neighborhood by their scratches and piles. Barney, a long-legged hound, has a deep, long scratch that invariably makes a V and digs up the turf. His feces are large and piled up. Jerry's piles are smaller and more strewn. His scratches are short, numerous, and shallow.

Some dog observers are pretty sure feces are placed by dogs as a pointed remark to their owner. Nele left her pile mark on a bed when her mistress scolded her harshly one day. Corky, a farm dog in Michigan, left his feces on the back-porch step for his master to see when pups (not his) born to the coonhound bitch got more attention than he did.

Rubbing parts of the body—head, neck, rump, side—on the ground and on other individuals is another kind of scent marking. All canids emit a personal scent, which can be released by fear but is more often offered up in affection. My Airedale would perfume the air with her "doggie smell," as people call it, when I came home after a long trip. She also emitted her scent when I was being particularly affectionate—holding her head in my hands, hugging her, kissing her nose. It was her way of saying what I was saying—"I love you."

Wolves and some dogs like the malamute have a gland on the top surface of the tail about four inches from the root. By rubbing their chin and cheeks on this gland and then rubbing their face against other members of the pack, which also contribute their gland perfumes, they mix individual scents to concoct a family odor. A wandering wolf knows when it is home by the ambrosia of its pack on the wind.

Rolling in (to us) foul-smelling scent is another olfactory joy of the canids. I always considered it a bad habit of an ill-trained dog until I was set straight by Michael Fox.

"Scent rolling is an esthetic thing," he said. "Dogs and wolves will roll in raunchy odors just so they can wear a different and exciting smell. It brings attention to them selves. All the other wolves and dogs come up to smell and investigate.

"Like your wearing a new dress," he added.

After Qimmiq came home one day smelling of dead fish, I asked trainer Ed Fouser if there was a way to stop a dog from rolling in scent. He shook his head—"Keep him on a leash."

## BODY LANGUAGE

When a person travels in a foreign land where he doesn't understand a word of the language, he can nevertheless communicate. He can see in a face whether he is to laugh at a joke or commiserate with sadness, approach in friendliness or be on guard against antagonism. And, knowing in general how human societies work, he isn't altogether baffled by what's going on around him. We may not be able to read canine gossip columns as they are posted on neighborhood fire hydrants, but by understanding social context and body language, we can guess our dog's feelings and be a better friend.

Your dog's social position is easy to see in its facial expression, posture, and movements. A dominant dog says it is boss by holding its head high like a monarch, its tail up and erect. It thrusts its chest forward and walks stiff legged, in almost a military stride. It may *ride up*—put a paw on the shoulder of a lesser-ranking individual to get its head above the other's and declare itself boss.

Dogs that are not as high ranking hold their heads and tails lower when they meet their superiors. Like the omega wolf, the lowest-ranking dog greets a more dominant dog with its head down and its tail between its legs. Its jaw sags and its eyes partially close. It is the embodiment of the phrase "hangdog look." When challenged, the omega dog rolls over on its back and surrenders.

The regal posture of dominant dogs, male or female, is accompanied by lordly facial expressions. Bosses wear smooth foreheads and wide-open eyes. Both the large eyes and the smooth expanse make the head appear larger. Their muzzles, too, are smooth, and their lips are relaxed.

Facial expressions change to signal a dog's mood and intentions, regardless of its social status. Therefore you can't rely on body posture alone; facial expression can modify the body's meaning. For example, the pose of the coward or dog who has been abused, is, at the sight of the enemy, back arched, rump pulled down and tucked under. But this is also the pose taken by a pup when asking an elder to play and a prelude to greeting man or dog friend when awakening from sleep. The face makes the difference. The coward holds its ears back and down, mouth closed, and eyes narrowed. The play-seeking pup is wide-eyed and smooth-faced. The awakening dog yawns as it arches.

The meaning of another gesture can only be guessed by context. The lion crouch, down on the belly with head held low to the ground like a lion stalking is intended to terrify an opponent when confronting it. The bitch also uses this pose when approaching her mate, and English shepherds will fall into the lion crouch when chasing balls or rounding up sheep. It expresses readiness for just about anything, and only the context can tell you what.

**THE DOG'S MAJOR FACIAL EXPRESSIONS**

| | |
|---|---|
| *Relaxed and confident* | *Ears back*<br>*Worried* |
| *Wrinkled brow*<br>*Lowered ears*<br>*Suspicious* | *Threat stare*<br>*Lips pulled back*<br>*Somewhat fearful* |
| *Fangs bared*<br>*Extremely fearful and threatening to bite* | *Apologetic* |

**HEADFIRST**

But there are facial expressions that carry specific meanings.

The full threat expression, "Get back!" is said with bared teeth, opened mouth with corners pulled forward, and a wrinkled head with ears erect and pointed forward.

Closed mouth with corners pulled far back, a smooth forehead with slitlike eyes, and ears drawn far back and close to the head is the face of the insecure—"I am afraid."

Suspicion is communicated by holding the ears out to the side and slightly down; the eyes are half-closed, the forehead creased between the eyes. The lips are raised only enough to slightly wrinkle the nose. The mouth is closed, and the corners are pulled downward.

The dog's emotional expressions of anxiety, happiness, peace of mind, anger, affection, jealousy, embarrassment, and hostility are not too different from our own ways of communicating our feelings by face and body. The ability of the dog to read our feelings seems like mental telepathy until we observe what we are saying with body and face.

My young friend Paul has a dog who lies down beside him when he is sick in bed with a cold or fever, but not when he is in bed relaxing or reading.

"When I'm sick, I look sick," Paul said after thinking about why his dog behaved as he did. "I droop and drag around, and Skip knows—just like I know when he's sick."

When the setter down the street got into the garbage, as he did from time to time, he would look at his mistress and sulk before she could open her mouth to reprimand him.

"My anger is written all over me," she said. "I finally realized my face talks to my dog. I scowl, I jut out my chin ever so slightly, clamp my lips together, and draw myself up when I'm mad at him.

"I don't need words. I've said them."

An embarrassed or guilty dog won't look you in the eye any more than will a child who knows he has done wrong. A wildly happy dog can't clamp its mouth glumly shut any more than you can when you're bursting with joy. What's more, glum and happy moods are catching. Laugh and your dog laughs with you, even though it doesn't get the joke.

## SUBTLETIES AND EMPHASES

The dog, in its adjustment to man, has evolved a few expressions especially for us. Lip licking means the dog is under stress and that possibly you and it are not communicating. You may be giving orders the dog does not understand.

Jaw tremors express intense emotion. My Airedale quivered her lower jaw when I brought another dog into the house, when she smelled a deer on the wind, when she heard a record of wolves howling, and when she looked upon a human infant.

Panting is another expression that has evolved, finally, to contact us. Originally it was only a device to expell heat from the body; then it was that plus a means of communicating to its own kind anticipated exertion. A pup that wants its parent to play runs in a circle with its rear end tucked under and its back arched. If the parent does not respond, the pup will stop, face the adult, and pant. It is not perspiring because it is overheated but because it is saying, "Play with me." Play generates heat, heat generates panting. What was originally an end, panting from exertion, has become a symbol for activity. It is used on us with more specific meanings. Spike, my father's hunting dog, panted when he saw him take down the gun from the rack in the living room. He was talking about the miles he would be covering on the trail of a fox or a deer. He was saying, as my father interpreted it, "Let's go hunting." Nele literally pants for cookies, even though that form of hunting requires hardly any exertion.

Such indirect communications between man and dog are most evident when it comes to going out in the car, and just about every dog owner speculates about what it is that sends his pet bolting to the door before he even picks up the keys. Like myself, many set up elaborate experiments to find out.

For years I thought it was the jingling of the car keys that sent the Airedale to the door. Once, I purposely left the keys in the car so that she would not hear them. The next day, I had no sooner decided to go to the store than she was at the door. I decided the signal was the putting on of my coat. The next day, I wore a sweater so that I would not have to put on a coat. When I was ready to go to the store, Jill got up and went to the door without a word or a whistle from me. I decided it must be the combing of my hair. I did not comb my hair the next day before going out. She was at the door. I thought it was the time of day. I did not take the car out for two days, although I opened the door and went into the garden. She did not rush to the door in excitement but followed me to the garden with loyal boredom.

Finally I began observing my movements and my facial expressions and realized I was telling her my plans with very subtle signals. I shuffle my feet a little bit when I have decided to go downtown in the car. I also leave what I am doing with a resolute movement as compared with the casual way I get up from the word processor or stop what I am doing when I am not going anywhere. But more important, I am definitely wearing my "I'm-going-to-get-in-the-car" pose and face—a straightening of the back and an absentminded open mouth as I think about what I have to do in town.

Of course, Jill took cues from my clothing, just as all dogs do. Putting on my mountain boots would send her to the door, sometimes hours before I was ready to go hiking. When I dressed in suit, high heels, and business coat, my straightened back and absentminded expression didn't fool her. My clothes said, "No dog." Sometimes she never bothered to lift her head from the rug when I went out the door in my city clothes.

We don't have the advantage of reading from a dog's

clothing what it has on its mind for that day, but shades of meaning are emphasized by a dog's face and body coloration. The dark nose, the light belly, dark trim on the ears and necks are somewhat akin to punctuation. The colors emphasize what is being said on the face and with the body. We use body punctuation also—lipstick, mustaches, jewelry, and hats. With them we emphasize our personalities or punctuate the message we want to get across. "I am sexually attractive" is one of the more obvious.

Dogs (with the exception of those bred for a solid coloration) have darker hair on the muzzle, forehead, around the eyes, along the rims and at the base of the ears, and on the tail. The inner lip is usually black. Pale hairs grow inside the ears and on the belly.

It was once thought that a pale underside simply meant a lack of pigmentation-stimulating sunshine on the underside of the animal, but no matter how it came to be, it is now in the language. The omega that flops in submission exposes its white fur—the white flag of surrender—to say, "I give up. Don't hurt me." Aggression is turned off as though by a switch; the aggressor stops its attack. No wolf can attack an individual that is flashing this signal of surrender, but, unhappily, the signal has descended in somewhat garbled form to dogs, and there are those that do not honor it.

Most do, even if the signaler is a one-color dog that lacks contrast between the belly and the rest of the body. The belly-up pose is sufficient. When Jill visited Sadie, an apricot-all-over poodle, Sadie rolled onto her back at the sight of one toothy display from Jill. Jill's face relaxed; she lost her desire to fight, turned, and trotted off.

Facial coloration emphasizes what is being said there. The dark hairs on the forehead gather together when the animal frowns, becoming darker as they are compressed into furrows. The displeasure is more clearly visible with this emphasis. The black inner lip contrasts sharply with the white canine teeth, so that when the lip is lifted, the cruel fangs draw the eye and inspire fear. The meaning stands out as if in quotes.

Black hairs on the dog's tail draw attention to that bold signaler, and raised hackles on the back are more prominent for their dark coloration, which contrasts with the lighter underfur. The inner ear has light-colored fur margined with darker fur—black or dark brown. The outlines make the expressive positions of the ears more easily seen.

## STREET TALK—BOSS TO BOSS

The best way to learn what your dog is saying with face and body is to take a walk with it in a neighborhood where dogs abound and see how it speaks to its own kind.

Dog conversations always begin with the greeting, which varies slightly depending on the status, sex, and the social relationship of the dogs. I took notes one day on the greeting between two boss males who knew each other.

The two came trotting up to one another. Both were confident, their heads and tails were held high, their foreheads were smooth as each brandished his alpha title. As they came together, they stiffened their front legs and lifted their heads and tails to get them higher than the other's. They were pretty equal.

Their eyes were round, their ears erect and partly forward. Suddenly they rushed together and stood head to tail, freely offering their rumps for an anal sniff. This was an act of confidence on both their parts. Insecure dogs keep their rear ends from being sniffed by lowering their tails and circling.

*Two boss dogs greeting*

The tension was building. I wondered who would give in first. The faces began to wrinkle. Creases appeared in the smooth foreheads. Confidence was eroding in both animals, but they were still equal. Simultaneously they lifted their lips to expose their wicked fangs, which gleamed against their black lips. They were both wearing the threat face, ready to attack. They wrinkled their noses, expressing anxiety. As they did so, they raised their hackles, making themselves appear larger, and growled. In dog talk the deeper the growl, the more dominant the individual, and each was trying to hit the basement of sound. This fact is useful when being firm with your dog. A deepened voice conveys your dominance in the situation.

The two boss dogs looked as if they were going to kill, and I stepped back a few feet.

The dogs circled, still flank to flank. Their foreheads were wrinkled both horizontally and vertically, the dark hairs on their faces making the lines more visible as they gathered in the grooves.

Posing alike, the bosses maintained the balance of power but were looking for a way out, "with honor," as they rolled their eyes to search for excuses to depart—a bird, a master, another dog.

Some subtle signals were exchanged that I couldn't detect, because, although they were still circling, they were stepping apart equally, as if by mutual agreement.

When they were ten feet apart, each, keeping an eye on the other, walked off, slowly at first, then faster and faster. They lifted their legs and peed simultaneously. Neither had to say, "I lose." Their faces relaxed as they walked off stiff legged, each saying he was boss dog.

**PAL TO PAL**

When two male friends meet, one of whom has already established his dominance, the greeting is much more relaxed. Major, a dominant dog, and Barney rushed to meet each other, tails wagging. Major sniffed nose, groin, and tail first. Then he permitted Barney to sniff his nose, groin, and tail while holding his head and tail higher than his friend's. There was no show of teeth, no contest. After gathering all the news about each other, they went off a short distance and urinated. That done, they put their signatures on the ground—deep claw marks made with front and back paws. Major and Barney then trotted off together down the lane.

Males do not fight females, and most will take outrageous abuse from a bitch with tolerance and forbearance. When Lucky, my dog Gunner's favorite mate (dogs are faithful if given a choice of mates), came to visit in the off-season, she would snap at Gunner, growl, and knock his feet out from under him. He replied by courteously wagging his tail and coming back for more.

**BITCH TO BITCH**

Two females, however, will tear into each other, particularly if they are terriers.

Jill, the Airedale terrier, and Koda, who was part terrier and who lived on the other side of my woods, were both fighters—and enemies. One encounter went like this:

Jill saw Koda approaching and ran to meet her as she came over the border into our yard. She permitted her to come about twenty feet onto our property, thus getting the advantage of having her enemy on her own turf. (Among birds and some mammals the individual fighting on its own property always wins.)

Koda dropped into the lion crouch and stared at Jill. Jill answered by keeping her face smooth, announcing the alpha she was (on her own turf), and asserted her dominance by pulling her ears forward. Koda crept forward on her belly for the kill. Jill flaunted her rear end to show just how confident and dominant she was. Koda responded by leaping to her feet. They rushed together and stood side by side, heads to tails. They growled, trying to bluff each other out. Minutes passed as they waited to see who would chicken out first. Jill took the offensive by showing not just one canine tooth but both. Her nose wrinkled. "Your life is not worth a bone chip," she was saying.

The talk moved to the ears and the corners of the mouth. Jill's ears stood up as far as she could get them, her mouth opened, and her lips lifted clear of all her front teeth as she put on the intense threat face. "I'm not afraid of you. Beware of me!"

Koda was intimidated by this show of power. She drew her ears slightly back. Her mouth pulled downward like a frightened child's. She snarled. Jill's ears went back for an instant in a quick admission of a lack of confidence in herself, then she gained control and shot them forward. She growled. Her growl was deeper than Koda's.

I walked toward the dogs hoping to prevent a fight. Jill

gave me the eye, asking me to do something before another bloody battle broke out between the two of them.

I had two choices: lunge at Koda, which might have triggered Jill to attack, or stand between them like a mother dog breaking up a puppy fight. I stepped between them, lifting my elbows to look bigger and growling my disapproval in my deepest voice to say in dog talk, "Knock it off." Koda ran back over the border. Jill sidled toward the house. As she walked, her rear was toward Koda, usually an expression of the retreating coward in dog talk. Safe on her own territory, Koda barked viciously. Distance changed the meaning of that retreat. Jill ignored her and trotted along beside me, head and tail up, announcing her victory.

## STRANGER TO STRANGER

When two strange dogs meet, do not know who is dominant, and therefore must find out in order to know how to relate to one another, the conversation might go something like this dog talk that I picked up on the main street of my hometown.

Two men were walking toward each other, one with a shepherd at his heels, the other a boxer. Neither dog was on a leash, but both were heeling nicely. About fifty feet from each other they broke away and rushed together. They sniffed noses. They held their tails out straight—horizontal to the ground. (This posture announces that they do not know who is boss. Watch for it and pull your dog away. It usually means that a fight to establish dominance is next.)

The owners did not seem to know what their dogs were saying, for they did not call them off. The dogs proceeded with the war talk. They presented rear ends simultaneously. (An inviolable dog law is that anus sniffing is truce time. No one attacks.) While these two gathered information about

*Greeting of strangers who don't know who's boss*

*Truce time*

*The fight begins*

each other, their owners came together, raised eyebrows, and simultaneously decided to leash their dogs.The shepherd's master took the leash out of his pocket, and the clip inadvertently hit the boxer in the leg. Thinking the shepherd had violated the truce and attacked him, the boxer went through the entire war vocabulary: wrinkled forehead, wrinkled nose, widened eyes emphasized further by dilated pupils, lips pulled back, two canines exposed, deep snarling growl, stiff legs. He attacked, grabbing the shepherd by the throat and twisting him to the ground. The shepherd, apparently a high-ranking dog, refused to roll onto his back in submission. He snarled as if to kill and flashed the whites of his eyes, a signal used by many mammals to admit being down but not beaten.

The masters grabbed their snarling, barking animals and pulled them apart. The shepherd was pulled into his nearby car, and the door was slammed shut. The incident seemed to be over.

The two men talked, probably about their dogs, and as they did, the boxer raised his leg and urinated on the pants cuff of the shepherd's master. Why? Was he making him familiar? Was he a new object in his environment? Or did he know what people think about such an action? Some dog talk is beyond me.

### THE BULLY

There is one final dog "greeting"—a brutal one—that puts both man and beast in their place: the *humiliation*, or *bully, strike*. There is a human counterpart, as I learned from a young lady my mother took me to meet as a child because she had such nice manners. Mother hoped I might learn graciousness from her. The little girl came slowly down the steps of her home holding out her skirt, curtsied, and said, "I'm

Mrs. Humphrey's little daughter." Hardly had I time to be impressed when she gave me a swift kick in the shins, reducing me to scum.

In the dog version the bully rushes a rival and hits him hard with his shoulder, then twists and strikes again with his rump. The victim yelps, the bully strikes again and again, reducing the rival to a bruised and hurting omega who cringes and creeps off in disgrace. This strike is also used by some of the large herding breeds to push cattle around and by wolves to throw their prey off balance. However used, the humiliation strike is a powerful message, but dogs are more honorable than "Mrs. Humphrey's little daughter." They warn you first.

Basil, a half Lab, half poodle belonging to my friend Ellan Young, gave me the humiliation strike one day as I approached her house. Bounding toward me, tail straight out and bristling, he hit me so hard with shoulder and rump that I cried out in pain and, yes, in humiliation. I should have known—his tail had broadcast his intentions.

### WATCH THAT TAIL

Ever since Basil told me off, I have learned to watch dog tails as signals of what is to come. The tail emphasizes what is going on up front. They are the ultimate semaphore. Siberian and Alaskan malamutes and the spitzes carry coiled dusters over their backs that signal over long distances. The Gordon setter slashes a baleenlike spear—slender to avoid brush and briars. The poodle sports a pompom on the end of its tail because breeders like the emphasis. And the Scottish terrier flourishes a vertical, tapered rod. Although tails are important signalers, pinschers, boxers, and those other dogs whose tails are clipped at birth do adapt to this handicap, possibly be-

| THE DOG'S TAIL POSITIONS | | |
|---|---|---|
| *"Who's boss?"* | *"I'm boss."* | *"I'm somebody important."* |
| *"I'm depressed."* | *"I'm a terrible dog."* | *"I'm an omega."* |
| *"I'm about to fight."* | *"Maybe I'll fight."* | *"I'm at ease, all's well."* |

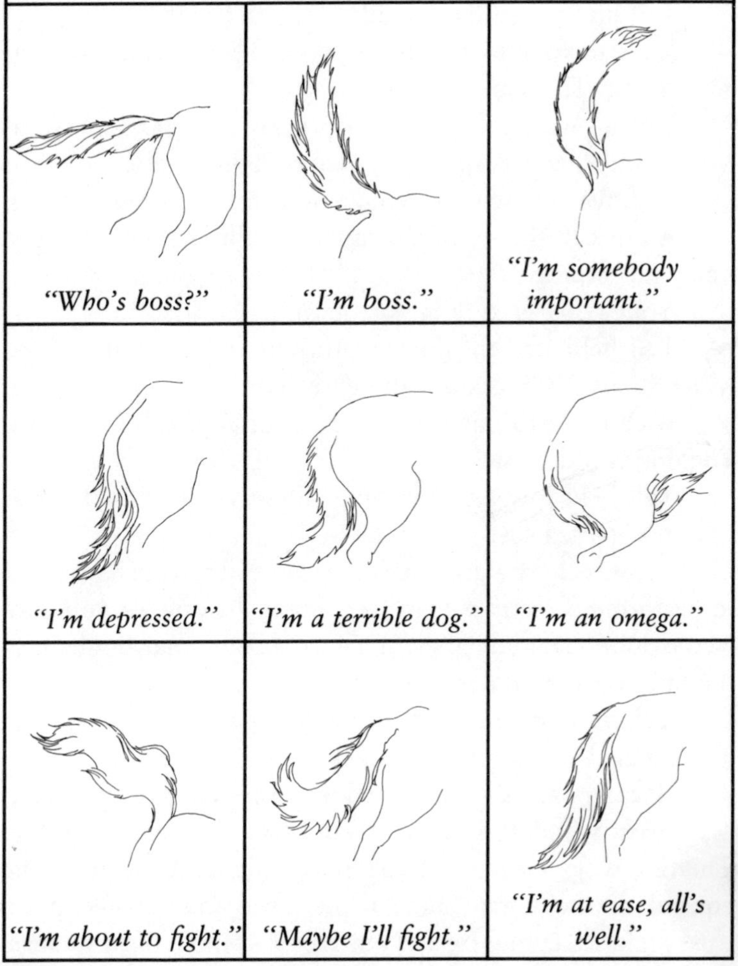

cause the tail is only a backup system for the face. The intense messages go on up front. Nevertheless, the tail should be watched for its own sake. It says a lot.

The tail has three properties: position, shape, and movement. The message, depending on the breed, goes something like this:

Tail up and gently curved: "I'm confident."

Your answer to this is to proceed with what you are both doing. The state of affairs is good.

Tail up with a crick (a sharp bend seen particularly in wolves and shepherds): "I'm going to fight."

Remove yourself and your dog from the opponent.

A crick in the tip of the tail: "I'm threatening you, but I may not attack."

Your reply is to dominate at all costs.

Tail held straight down with hairs bristling at the tip, which is slightly raised: "I'm depressed."

An answer to this is more attention or a call to the vet. The dog could be sick.

Tail held low and brushing sideways, no raised hairs: "I'm no good. I cast myself down."

Answer this by discerning the cause. Too much pressure to perform correctly is being put upon the dog, or you may be too dominant and harsh for a particular animal. Sometimes the tail says this to another dog.

Tail straight out and all hairs bristling: "Don't come any closer. I warn you. I'll attack."

Back up, as I should have done to escape Basil the bully.

Most wonderful, the dog's tail wag is its tribute to life. The dog wags its tail only at living things. A tail wag, the equivalent of a human smile, is bestowed upon people, dogs, cats, squirrels, even mice and butterflies—but no lifeless things. A dog won't wag its tail to its dinner or to a bed, car, stick, or even a bone. The dog appreciates and responds to aliveness; it is not a materialist, and we love that quality in it. Joshua, my friend Sara Stein's child, upon noting the power of expression in his dog's tail, asked for a fake tail when he was little but mourned that he couldn't wag it and talk like a dog talks.

**HOW TO TALK IN DOG POSES**

To some extent, you can mimic dog body language to say some things to your pet if you loosen up enough and do not mind looking ridiculous. Get down on all fours and spank the ground to ask your dog to play, or bow slightly and slap your legs. Grasp its muzzle in your mouth to assert your leadership, or, if that seems primitive, lay your chin over its muzzle, or take its nose in your hand and shake its head gently back and forth—same message. And, if you have an omega that cowers at your step, roll on your back to say that it is no worse than you.

My son Craig used this body talk on Sadie, who belonged to his best friend, Tom Melvin. Sadie and Craig had a problem. When Craig came to visit, which was often, Sadie would cringe, roll over on her back, and, using the most self-demeaning statement in all of dogdom, urinate in that position. She retrogressed to a helpless pup.

This was, needless to say, a problem. Papers were put out and rags were made available for Sadie's greeting to Craig. Visits were curtailed. Then Craig came up with an inspiration after reading about the language of the wolves. He opened the door to Tom's house one day, called Sadie, and before she could roll over, went down on all fours and rolled onto his back, hands folded over his chest, knees pulled up. Sadie braked to a halt, sniffed Craig's face, and put a paw on his chest to ride up in dominance. Thereafter, she met him head-on with her tail and head up. The rags and newspapers were put away.

**A TIME FOR LOVE**

A charming characteristic of the wolves that some dogs have is the habit of bringing home presents to the family. I watched the beta wolf of Sanctuary River bring a stout white stick to the den area and drop it in the midst of the pups. They pounced on it, carried it around, and fought for it, recalling to me Sara, the dog of my children's youth.

Sara, half Lab, half Newfoundland, often came home with the kids' jackets, shoes, boots, hats, even school books that had been laid on the ground while the children played. Once, she brought home a turkey left on the neighbor's back porch by a delivery man. On my daughter Twig's fourteenth birthday, Sara stood at the door with a bouquet of asters wrapped in florist's paper. Phone calls to the neighbors and local florists turned up no owner, so we put the flowers in a vase and enjoyed the gift. To this day, my daughter insists that the quiver of celebration in the air stirred the ancient wolf-giver in Sara, and so she went out and got a present.

Siberian and Alaskan malamutes are also generous gift bringers. Qimmiq had a strong thought in his head when he bit off a small branch of rhododendron and carried it up on the porch. It lay there until I came to the door. Then he picked it up, put it in my palm, and let go, making no effort to take it back as he does with his play sticks. I had been given a present, and I thanked him with effusive palaver and a hug. To kiss him on the nose, which I do on some occasions, would have been an insult in this instance. The body language would have been pulling rank, and Qimmiq was already doing what he thought was appropriate.

## A COMMON TONGUE

Finally, the dog communicates with us and its kind by voice, the medium we understand best, although it is not the dog's most expressive vehicle. The dog has five sounds: the *whimper* (also called the *whine*), the *yelp*, the *growl*, the *bark*, and, in some breeds, the *howl*.

The whimper is a group of high, plaintive cries, softly uttered, that sound pitiful and helpless. It is used by puppies to get their mother's attention and by adult dogs to get their master's attention and to beg for something they want.

When the whimper crescendoes into a sound so high-pitched and so penetrating that it seems to pierce the eardrum, it is called a whine. Sounded higher, the whine goes out of range of the human ear, and only the face and eyes reveal that anything is being said. A slight frown is visible, and the eyes are pulled down slightly at the corners to invite concern and, it is hoped, affection. Females whimper and whine more than males, for these pleas are the language of the estrous season, spoken to invite courtship. My dog Jill whimpered in the middle of the night to ask me to open the door and let her run with her lovers. She also directed her whimpers to me, saying, "I want to be your friend." The same sound is also called the social squeak; pet wolves make it to invite friendliness with people or dogs.

If you talk to your dog with the whimper, you might get a baffled look with the head tilted to one side. After all, alphas don't whimper for attention. Past that initial puzzled look, which is actually tuning in the ears better to figure out your terrible accent, you might get a tail wag for an answer. Quite often a dog will give whimpering people the "Let's-play" response, spanking the ground with the front paws, holding the rear end up.

The yelp is agony. It turns the heads of masters and dogs and needs no translation. Even a child reacts to the pain in the cry and hurries to rescue the dog. The yelp has shades of meaning depending on the age of the dog. The puppy yelp, as mentioned earlier, means surprise, hunger, pain, and cold; a yelp from an adult dog says it is injured or thinks it is about to be injured. The adult also yelps from psychological injury when given the humiliation blow.

The growl is a low, muttered complaint, and its meaning is eminently clear. Dogs growl when they are threatening one another while standing head to tail, when they warn off a stranger, when they are protecting house, den, or food, or when they want to be left alone. The deeper the growl, as with Jill and Koda, the more dominant the dog. It is an unfriendly statement.

I have only known one person who could put off a dog by using a growl. The postman who delivered our mail in Poughkeepsie, New York, could send our hound into full retreat by growling deep in his throat as he came up the walk. Yet, oddly enough, sound experts claim that the sound-wave pattern of human word whining ("Please, Mommy, pretty please") and human growling ("Get out of here!") are the same as those of dog and wolf whines and growls.

The bark of the dog is one of the least pleasant of animal vocalizations. It is monotonous and boring, a single note usually repeated over and over again to say, "Someone is approaching the house," or "Stand back, stranger," or "A dog is on our property," or just "Here I am." A professor of animal behavior at the University of Chicago counted the barks of the champion of barkers, the cocker spaniel: 907 barks in ten minutes. My uncle had a cocker that could run that dog a good second, and I am sorry to say no one was unhappy when he died.

Dogs, like wolves, have two kinds of bark. One is alarm, the other is a threat or a challenge to intruders. The alarm bark is short. The threatening or challenging bark is often several sharp barks followed by more drawn-out ones. Directed toward you, either one can be a demand to be fed or to go on an outing, depending on the circumstances.

At night the challenge bark becomes a roster. When the sounds of traffic die down and the last train has rumbled off to the city, Leon, a mutt at the bottom of the hill near the swamp, awakens and barks once. I hear him and wait. He barks again, then again; and he waits. Apparently he hears another voice that I cannot, for the next time Leon barks, he is excited. Finally I hear the distant dog, for someone has let him out. He barks, "*Bow, bow, bow,*" for almost two minutes. A third dog speaks up, then far over the hill toward the next town a fourth sounds. The dogs are checking in, calling out their identities, letting each other know where they are, keeping tabs, keeping in touch—messages they can send only when the town has quieted down and all but the dogs and night animals are asleep. I find the night check-in reassuring. The dogs, by taking roll call, are saying, "All is well."

In contrast to the bark, the howl is a colorful and thrilling song. It is a gift from the wolf passed down through four thousand generations to some domestic dogs.

## THE HOWL OF THE WILD

Dog howlers are the malamutes, huskies, German shepherds, and individuals of some other breeds. Some individuals howl only as pups, which suggests that the urge to howl has been weakened in breeding or that it is stifled with age. A few dogs that aren't ordinarily howlers will howl if they hear howling. When I played the phonograph record *The Language and Music of the Wolves*, sung by wolves and narrated by Robert Redford, Jill listened in great anxiety and said so by trotting restlessly around the room, whimpering, licking my hand, and trembling her jaw. Eventually she sat down, formed her mouth into an O, and howled with the wolves. As she joined the electronic chorus, she seemed to lose all contact with her human family and slip off to some distant landscape.

Scientists are still unsure why dogs howl, but just about every owner has his opinion. Ramon, a Baja fisherman, says his dog howls when he is lonely for the wilderness. Hank, an artist who lives in the next town, thinks his shepherd howls because his ears hurt when the fire whistle blares. The violin

teacher at the bottom of the hill believes his dog howls to tell a student he has hit a sour note.

A dog might howl for the same reason the wolf howls: to bind the pack together before the hunt. Wolves howl also to identify themselves to their fellows when scouting separately for food or trotting off to check a border alone. Mech's latest research indicates that the wolf pack howls to space themselves and keep distance between the packs.

Michael Fox speculates that howls, both solo and group, might be brought on by restless anxiety: The prey is not in sight, the pups are hungry, a storm is coming.

Many of us who have heard wolves howl feel it is also a celebration.The music is rollicking, at times jubilant, and the tails wag joyfully.

The group howl, like our musical events, has form. The sing-along is generally initiated by the alpha male wolf with a series of short tuneful notes. The second wolf, either the beta male or the alpha female, comes in with a solo, the others join one at a time, then all howl in unison, in expectation, in magic. The howls start low and work up to shorter and higher notes that break into barks and end abruptly when the breath is used up.

Wolves appreciate people joining their choruses, but they have to obey their rules. Each wolf wants its own note. Anyone who joins in must harmonize. When Mech was howling with the wolves of Ely one moonlit night, he howled the same note as one of the pack. The wolf stopped, listened to Mech's note, then politely took another.

In questioning the experts about the meaning of howling, it was Gordon Haber who put it most succinctly: "It's fun." And it is. Upon the conclusion of a wolf-cry musical composition at the Kennedy Center for the Performing Arts in 1979, composer-conductor Paul Winter turned to the audience. "And now," he said, lifting his baton, "let's all howl. It is wonderful to do." We threw back our heads, rounded our mouths, and, closing our eyes, howled as if we had finally found the joyful road to the center of the universe.

**PLAIN ENGLISH**

Mostly, dogs are very much in our world. Another "How-does-my-dog-know-when-I-am-going-out?" story makes the point. The clue that Groschen, a German shepherd, used baffled her master for years.

"I would be sitting at the breakfast table with my wife, still drinking my coffee and talking about my plans," my friend Dave Pollack told me, "when suddenly Groschen would get up and go to the door. I was going to drive to town, and she knew it, although I never made a move.

"For years I experimented with all the gimmicks that might have tipped her off. I jangled the car keys, put on my coat, took out my wallet when I was not going to town. But I didn't fool her; she didn't go to the door.

"One day my wife was sick. I carried a cup of coffee upstairs to her. The dog was sleeping on the floor beside the bed.

" 'I'm going to town,' I said, as I always do before departing. 'What do you need?'

"Groschen walked out of the room and went downstairs to the door.

" 'She understands English!' I exclaimed.

"My wife looked at me in disgust.

" 'Of course, she does! What do you think she understands—German?' "

Dogs do learn our language. When my son and I were in Baja, Mexico, he called, "Here, doggie!" to a black-and-white dog trotting across the street of Cabo San Lucas. The dog trotted on.

"Perro, aquí!" he tried. Instantly the little dog turned around and came running up to him. The message was not in the tone of Craig's voice or in his body language. The dog knew Spanish.

Dogs are much better at learning our language than we are at learning theirs. Statements like "Let's go for a walk" are understood by most dogs even when said casually. A neighbor of mine looks at her dog around nine o'clock in the evening and says, "Shanty, it's time to go to bed." The bright-eyed mutt looks up at her, arises, and trots off to the cellar where his bed is laid.

And that smart Nele goes into conniptions when someone mentions the word "cookie." The family has been forced to spell out the word to prevent Nele from barging into the kitchen and making a scene.

One friend counted twenty-seven words that his dog knew and responded to, "and many more," he added, "when you consider that I put those key words in phrases."

A dog is more than man's best friend and more than a faithful Old Dog Tray. The very nature of its sensory system, its feelings for family, and its love of life make it, for better or for worse, like a good, loving partner in marriage. After a few years of talking and gesturing to your dog, the two of you come into exquisite communication, as in those close marriages in which the spouses know each other so well that their anticipation of one another's meanings is the closest thing we know to mind reading. Qimmiq need not whine for a biscuit anymore—he glances from me to the closet where I keep them. "Okay," I say, open the door, and pick one up. Sometimes I say nothing, just act. In the woods I need not ask him to sit when we come to the top of the hill in view of the glorious Hudson. He glances at me, then the vista, sits down, and, like myself, gazes across the river valley.

Only a few weeks ago we were on a new trail that opened up over a lake. Qimmiq glanced back at me, ran to the ridge, and sat down. "You're right, it is beautiful," I said. He wagged his tail. His wild kin, the wolves of Mount McKinley, dig their dens high on hills in view of gray-green valleys and snow-covered peaks. And they, like Qimmiq and me, sit and enjoy the magnificence. At such moments a glance from either of us will say a volume, and the abyss between species is crossed from both sides.

But the course of talking to your dog does not always run that smoothly.

I was panting to Qimmiq, asking him in his own language to play with me, but he did not respond. I was persistent—hanging out my tongue, hyperventilating and dreaming as I did so of that different land from which he had come, the Arctic. It is a land of the northern lights, where the moon sets and rises an hour later, of sun dogs and ice fog, of hoarfrost and no rats, mice, cockroaches, or fleas. Carried away by my daydreams, I spoke in my own tongue.

"And so, beautiful dog," said I out loud, "is it that land that makes you so noble? Is it that sky that makes you so beautiful?"

He frowned and, wishing to please, went upstairs and brought down my sneaker.

# The Cat

The cat Danny was not fed for the two days that my friend Joan Gordon was delayed out of town. He had a door through which he could come and go to his hunting grounds, where he often caught mice, and Joan was not overly concerned about him. He was, after all, a cat, capable and independent, the perfect predator. Joan had left him on his own before.

As she came up the path to the house, she smiled to hear Danny meowing his chirruping welcome. Eagerly she opened the door.

"Hello, Danny, old fellow. I'm glad to see you. Hello." The hefty, ruddy tabby cat looked right at her face and chirruped again. His fur was pressed lightly to his body, his whiskers were bowed forward, his pupils were dilated with pleasure, his tail was held straight up like a flag pole, and he danced on his toes: an altogether exuberant greeting, perhaps best translated as "Hello, hello, hello, hello."

But something amiss caught Joan's attention. A kitchen

chair had been knocked over, by Danny no doubt, and one leg had jammed his door closed. The cat had not eaten after all.

With great concern Joan dropped her coat and suitcase, opened a can of food, and put it on the floor.

But Danny did not eat.

Hungry as he must have been, he ran to Joan and repeated the redundant greeting. Next he rubbed Joan's ankles with his head, then with his flank, and then snaked his tail over her shins, all the while purring. He arched his back toward her hand, asking to be petted. She obliged and then again urged food on him.

Danny ignored the food and went on with his cat talk. He rose on his hind legs and arched his shoulders and neck toward Joan's hand, asking more forcefully this time to be petted again. When Joan stroked him, he purred like a motorcycle. Finally Danny turned to the food, only to take one bite and return to repeat the entire exuberant, sensual, deeply felt, and minutes-long "Welcome-home" routine again.

Joan called me that night, incredulous. Danny, she now knew, put her before a can of food.

## A TALK WITH AN UNEXPECTED GUEST

It is quite typical of cats to greet their human friends before eating—even after a prolonged fast. Some insist on socializing as a prelude to any meal. Skitters, the plume-tailed coon cat of my doctor's wife, will not touch the food Marilyn has put down each morning until he has sniffed her and rubbed her with his flanks. He arches, asking to be stroked. Marilyn caresses him, and Skitters then will take a bite, but still he will not finish his food until she has picked him up and performed the ultimate in cat greetings—best described as head bumping.

Cats recognize humans, faces and voices, and even their cars. One or two minutes before Joan pulls into her driveway, Danny hears her car approaching many blocks away and stations himself at the door to greet her. Not only do cats actively seek the affection of their "owner," they tolerate a

*Head bumping*

good deal of abuse from other members of their family. Children who, for some reason, are wont to pick up a kitty by the middle, causing it to dangle head down, are less often scratched than they deserve, and some remarkably pacific cats allow themselves to be dressed in baby clothes and pushed in a pram. Not a few cats consider that the only appropriate sleeping arrangement is to be cuddled in bed with their human. Paul Leyhausen, a German scientist who studied house cats for twenty years in great thoroughness and detail, concluded in no uncertain terms that "the domestic cat can have a warm and loving relationship with human beings."

Yet the friendship between humans and cats is something that scientists would not have predicted and that they find difficult to explain. Unlike dogs, which are social animals and respond to our family as they respond to their own, cats are loners. They scorn other cats, aggressively keeping them at a distance or passively avoiding meetings whenever possible. They hunt by themselves.

Cats are designed for a pounce-from-ambush hunting strategy. "Wild" colors and patterns, still seen in the tawny shades of the domesticated Abyssinian breed and the familiar red- or gray-striped tabbies, are camouflage that can render the immobile cat invisible as it silently awaits its prey. The feline backbone is exceptionally limber; it has five more sacral vertebrae than the human backbone, enabling cats to flex, stretch, and twist a good 180 degrees. Although not cut out for marathons, cats in full stride can cover four times the length of their body at each leap, change direction in midair, and pounce upon their prey with deadly speed and agility.

Their eyes are larger than any other carnivore's in relation to their head size, a feature that attracts people because it gives them a babylike and innocent look. Their excellent sight is surpassed only by their exquisite sense of smell and by hearing that is deftly tuned to the rustling sounds of prey in dry grass or leaves and the ultrasonic squeaks of mice. Cats are the most successful of the mammal predators, catching their quarry 50 percent of the time as compared with wolves' 20 percent, and a good mouser of the pussycat sort does even better.

Wild kittens, like the young of most mammals, are friendly and trusting of humans. But the wild cats of the world cannot form reliably warm relationships with people that last into adulthood. They can be "tamed," even trained, as circus animals, for instance; they cannot be relieved of an inherent fear of man. What friendliness can be achieved by an individual cat is not carried on into the next generation. A British zoologist raised from kittens two wild Egyptian, or Kaffir, cats, which are considered to be the progenitors of the domestic cat. They were tame and pleasant creatures, but their offspring were as wary of humans as if they had grown up in the wild. They ran from people, stole food from the table, and when smacked for doing so, would not jump down as do domestic cats but would stand their ground, lay back their ears, spit, bare their teeth, and try to bite.

Dogs can be bred back to wolves or coyotes and still remain domestic animals. That also is not reliably so in cats. Naturalist Frances Pitt of England reported that the offspring of a mating between the Kaffir's wild European relative and a domestic cat had inherited the disposition of the wild cat. They were nervous and bad tempered, attacked Pitt's geese, and ran off to the woods. She had to lock them up because they could not be trusted.

So why is it that house cats live so well with people?

Geneticists offer an explanation: Domestication was enabled by a genetic change, but a very subtle and tenuous one. Although the domestic cat, *Felis catus*, resembles its wild counterpart, *Felis sylvestris,* in almost every other respect, one thing is different: It fails to develop wariness toward people in adulthood.

The change was not something accomplished by people who captured, tamed, and bred wild cats to a more relaxed disposition, but people did have something to do with it. Sometime before about seven thousand years ago, it seems likely, wild Kaffir cats in North Africa and the Middle East began to take advantage of human civilizations then developing along the Nile and other rivers. The people there were forming civilizations centering on the growing and storing of grain. This massive collection of seeds permitted mice to thrive on a scale that no wild rodent had before enjoyed and that no rodent predator had before encountered. The wild cats moved in close to this bonanza. These oddball cats that could tolerate people lived well and passed on this boldness to their offspring. So one could say that while people provided the opportunity, cats pretty much domesticated themselves. Their evolutionary "guess" happened to be correct, for in fact we did not do battle with cats but welcomed their predations on destructive rodents. Consequently, this charming beast—which does not defer to a leader as dogs do, is not a follower like horses, sheep, and cattle, and is not gregarious like chickens and ducks—found us to be pleasant enough, and our supply of rodents rich enough, to move intimately into our lives by perhaps 4000 B.C. in Egypt. Cat remains from that time show the first signs that some were being purposefully bred, perhaps as pets.

The human-tolerant cat seems to have lived within our communities for more than a millennium before we took historical note of its presence or expressed our admiration for its talents and beauty. But by 2500 B.C. the cat's likeness was being drawn on Egyptian papyrus, painted on frescoes, shaped in silver and gold, sculptured in ivory, and carved out of precious gems.

Around 2000 B.C. the cat was elevated to the status of a goddess—Bast, symbol of grace, fertility, and femininity. So revered was this animal that it became a felony to kill it. Bodies of cats were mummified with the reverence and care accorded a pharaoh; some were buried in gold caskets. They must have been, by then, beloved pets. Yet most were still physically identical to Kaffir cats, *Felis sylvestris*, not *Felis catus*. Mummified cats found in the tombs of pharaohs of 2000 B.C. are Kaffir cats, according to zoologists of the London Museum of Natural History, and all but three of the 300,000 mummified cats entombed around 1800 B.C. at the village of Bani Hasan, are also Kaffirs. Even by 1350 B.C., when Queen Nefertiti, the beauty of antiquity, painted her eyes in the likeness of cats' eyes, the pattern she chose to honor resembles the facial markings of the Kaffir.

Ultimately the domestic cat was distinguished from its wilder ancestors by a slightly altered body, especially by shorter legs and a narrower skull, but the differences are not nearly so dramatic as are those between domestic dogs and their canid ancestors.

Exportation of the cat from Egypt was forbidden during the years of its deification, so the domestic cat did not appear elsewhere until around 1000 A.D., when it was brought to China. Several centuries later, cats spread to Japan and Europe, possibly making their way to these countries smuggled aboard the ships of the globe-traveling Phoenicians, who gave no credence to the Egyptian belief in feline sacredness. Once scattered, these pleasant cats mated with various closely related wild cats of Europe and Asia, whose genes are now also represented in the many modern breeds. Although there are at least thirty breeds of cats today, most are not unlike the original. Black (melanistic) fur, white (albinistic) fur, and fur with Siamese coloration are the result of mutations, as are the tailless rear end of the Manx and coats of odd textures, such as the curly coat of the rex.

In Europe during the Middle Ages the cat was shunned as a symbol of paganism. It was considered an animal of the devil and killed in great numbers. Only a hint of that dread still lingers in superstitions and Halloween traditions, but the cat remains the most controversial of all domestic animals. There are cat lovers, and cat haters, and few in-betweeners.

The fact is that the genetic change that launched cats toward domesticity left them surprisingly unchanged. House cats have retained a good measure of wildness; some people like that, and some people do not. Their fights are passionate, their pleasures sensuous; yet their killing is a model of cool precision, and they cannot hide a fundamental aloofness that stems from their solitary way of life.

Cats can and do live on their own without help from people. Feral cats and kittens raised in woods, parks, or city streets spit and press back their ears like their wild brethren when they meet up with people. Even pet cats are not totally dependent upon us, and we have not much altered their appearance or behavior with breeding. Their hunting skills have not changed. Their favored prey is mice and rats. They still don't care much for the company of other cats. After all these millennia, the cat remains an unexpected guest, who can be convinced to stay only by nurturing its subtle failure of wariness and by courting it on its own wild terms. Some people welcome the challenge. Others close the door in its face.

### ON BEING A FAVORED PLACE

Whereas a dog joins in with us, a cat moves in. Both kittens and old cats take over the whole environment, which includes mice in the woodwork, pathways through the yard, food in a dish, sunning spots, lookout posts, hideouts, resting places, and, as it happens, humans.

Upon moving in, the cat brings with it from the wilds the cat concept of home. Immediately it cases the area and, whether it is a newly weaned baby or an adopted adult, sets up a uniquely catlike kind of territory divided into a central *first-order home* and a surrounding *second-order* area. The first-order home may correspond to your home, or to only part of it, or to home plus yard. The second-order home corresponds roughly to our idea of neighborhood. What makes the central and peripheral areas unlike wolf or human territory is that the cat does not seem to consider them whole, inviolable spaces. Rather, your home to your cat is a number

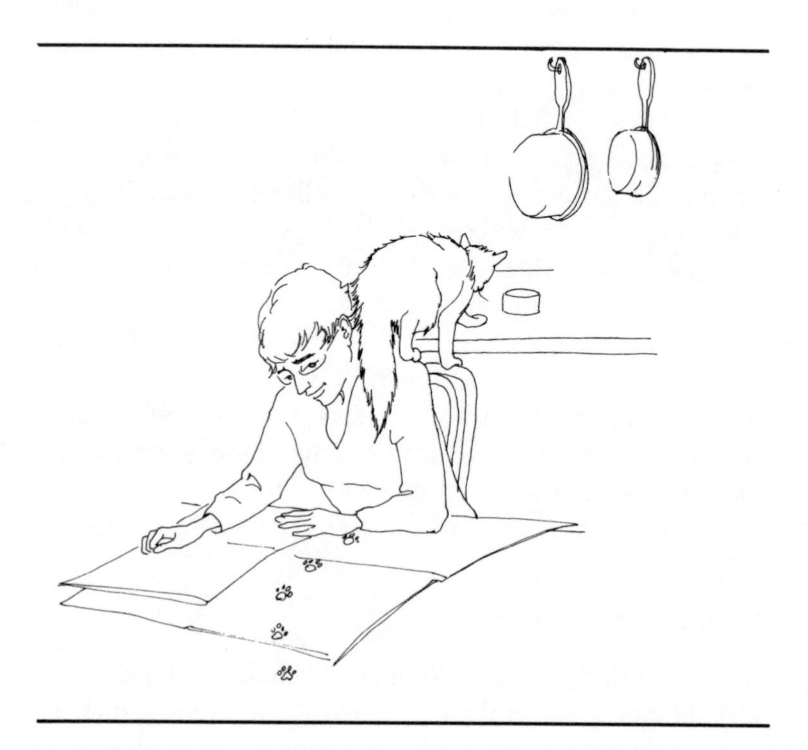

of regularly visited localities—places for resting, sunbathing, keeping watch, and eating. These are connected by an elaborate network of pathways. Although the cat adheres to the paths, it knows all the territory in between and scrutinizes it daily. But the route and the favorite places it connects are the thing. My friend Sara Stein's huge black-and-white cat, William, meows to clear her off his path to his morning sunbathing stop but does not bother with formalities on his route to his breakfast place, which is over the kitchen table and down. If she is sitting there, it is onto her shoulder and down her back.

After breakfast, William retraces that path to his first rest stop, Sara's lap. Although he purrs and enjoys being petted, it is the place that matters. If Sara is concentrating on the newspaper and not moving to acknowledge his presence, William "arranges" her arm to suit by pulling it over with both front paws. Then he lays his head down upon it. He snoozes, purring just as loudly as he would if he were being lavishly petted. He is happiest, Sara has learned by making mistakes, when his "place" does not cross her legs, shift her position, or turn the pages of the paper.

Although we make laps all the time, it is your cat's decision as to when to use them. You can't dictate to a cat. My daughter Twig had a charming tortoiseshell female named Trinket. She lived and raised her annual litter of kittens with us for thirteen years. Yet whenever I wanted Trinket in my lap, she would not stay more than a moment. When she decided it was lap time, I could not get rid of her.

Sara says William's routine was established when he was a tiny kitten, one of a litter raised in her home. Cats are very much creatures of habit who expect their environment to produce laps, patches of sunlight, and bowls of food on schedule. Making your habits and your person fit your cat's expectations are matters of noticing its daily appointment calendar and of seeing to it that the routine goes smoothly. In that way you are not only a favorite spot within its first-order home but also its reliable activity director and meal planner. That's the secret behind talking to a cat. Silly, the cat of an elderly couple who themselves disliked disruption in their daily routines, was from kittenhood enticed into the basement for the night with an evening meal and the whistled tune of "Danny Boy." The tune eventually sent Silly to the basement in anticipation of a meal whenever the couple wished.

Ordering the cat to the basement wouldn't have worked. Cats have no respect for authority. They are not cooperative and friendly pack animals. Give a cat a stern command and you are likely to be ignored or swatted. Put a choke collar on a cat to discipline it and you have a cat on its back with its claws and teeth raking your arm.

Yet house cats can learn all sorts of tricks if their environment is forthcoming and reliable. Ernie Dickinson, a friend and writer, has a stunning yellow cat named Georgie Pillson who, upon request, leaps through hoops, jumps from the floor into his arms, rolls over, and, upon hearing the jingle of car keys, runs to the door for a ride. Georgie walks in the woods with Ernie without a leash and shakes hands with strangers. He is now learning how to walk a tightrope and how to press his paw on an ink pad in preparation for signing this book when it is published. Ernie did not train Georgie by asserting his authority. He wooed him patiently, with bits of chicken liver.

"Cats are trained by rewarding them with food," Ernie said. "But once a trick is learned, praise and attention suffice. Georgie now rolls over for me without any reward except my applause and verbal praise. For that he will do it again and again."

However, Ernie admits that there are limitations to being a responsive environment. Georgie Pillson will not use the carpeted wooden cave Ernie made for him or a sleeping bag he gave to him, preferring three velvet chairs and the real beds. The house is, after all, his oddly piecemeal first-order environment, and if he chooses to ignore some aspect of it, that portion is not "there" to him.

Even professionals can't do much about the ultimate independence of the best-trained cats. A house cat who performed in a small traveling circus was so acrobatic that my son Craig bought me a ticket to his performance. He could hardly wait for my return to share with me his delight in this cat's astonishing skills.

"What's so great about a cat sleeping on the top of a stepladder?" I asked when I got home.

"Well, that's confusing," Craig said. "The cat leaped through hoops and turned flips in the air when I saw him last night."

A cat is like that. It is not a hail-fellow-well-met who will go along with any nutty activity its host suggests, as will a dog. Its attitude is that of a relative settling in, expecting concessions to be made on both sides.

## WILL YOU BE MINE?

Like a relative who has come to stay, a cat soon makes itself at home by leaving its personal "belongings" around. Luckily, this is not still more empty glasses and old magazines to contend with but invisible and, to us, unsmellable pheromones, the cat's personal odor. Cats produce these pheromones in glands discovered by British scientist R. G. Prescott, who discerned their locations by noting those spots where cats touch and rub people, furniture, trees, and other "beloved

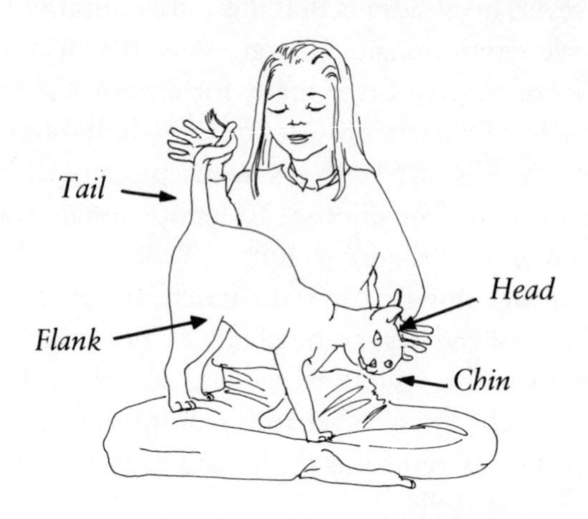


*The locations of pheromone glands*

objects." The scent glands are in the head, along the lips, beneath the chin, and strung along the flank and tail.

Those odoriferous areas cats rub liberally on everything from people to posts. When they lace us with these friendly and intimate odors, they are labeling and stabilizing us in their environment. They perfume us with grace and often with contented purrs—as when Joan and Danny bumped heads. Because the touch seems loving, and often is, we tend to ignore the real point of the message. Pheromone rubbing is not so much an outgoing confession of love as it is the stamping of property with ownership labels.

But the more responsive you are to being marked as your cat's own object by stroking and rubbing in the right places, the more heavily you will be scented and the more beloved you will be. Those who avoid a cat's marking are flatly stating that they refuse to belong.

Cats can be quite miffed by such a rejection from their

environment and then all the more aggressively insist on having things their own way. Cats are wizards at sensing which people are trying to avoid them, and they rub pheromones on them ad nauseam until picked up and tossed away. Every time I walk into the apartment where Freud, a black cat with a white star on his chest, lives, I am rubbed to exasperation. If I stand and talk with his master, a psychiatrist friend of mine, Freud is around my ankles marking me with his chin, and if I sit down, he is on my lap rubbing his head and tail against me. I am forced to either throw him to the ground or spit at him in his own language, my teeth bared. Chastised, he waits under a chair until I am involved in conversation, then returns to harass me with pheromones. I never have achieved a decent relationship with that cat.

Cats also present their scent glands to other cats during courtship, to their young when mothering, and to siblings and mother when they are kittens. Scents thus serve as a bonding language in genuine social relationships, as well as a personal stamp on belongings. You can talk love and affection to your cat through these gland points even when the cat is not the initiator. Stroke the head, chin, flank, tail—you will be answered with a possessive purr.

## SAY IT WITH A STARE

Beyond the cat's first-order home is the second-order one, less beloved but no less important to free-ranging cats. To a wild cat, the second-order home is the wider area that surrounds its central, best-loved, and most-frequented place. To the house cat it is beyond the house and garden, out in the surrounding yards, woods, and fields. You could call it the work area, for the second-order home is a network of trails that lead to grounds for hunting, courting, fighting, and other cat business. Cats who are protected from the dangers of such adventures by being kept indoors do not have second-order homes.

The outer limits of second-order trails for free-roaming cats can be an area half to one square kilometer for a female and twice that for an unaltered male. Just as cats have a sense of territory somewhat different from our concept of a specific area with an invisible fence drawn around it, so they do not share our attitude about protecting property, as do wolves and many other mammals. They don't mark boundaries and don't defend the area at all costs against others of the same species. As far as cats are concerned, all parts of their grounds (except for the den where babies are being raised) are there to be used by other cats on occasion, provided they occupy path or place one at a time and under the right conditions. The trouble is, the "conditions" are strictly personal. Trinket let one of the neighborhood cats sit on her favorite pillow in her first-order home—but only that cat and only when she wasn't occupying it.

All of the trails through second-order territory make sense. Their destinations are important to cat life. People who let their cats out of the house at night and think of them wandering the country on wild and nefarious expeditions are wrong. Cats have exact destinations in mind, take precise routes to these places, and return along them when the mission is completed. If the mission is mating, toms might not come home for a week or more. But cats are not lost. They know where they are and why they are there, and, unless hurt or "adopted" by others, they will return home. The cats that "come in out of the cold" and adopt you are usually cats that have been dropped along a street or road side and abandoned. They do not leave home and choose to live elsewhere unless they have been abused by other cats, dogs, or humans. Cats love their places and their pathways.

My backyard, on a wooded hillside in Chappaqua, New York, is laced with cat pathways originally laid down by Trinket and now used and sustained since her death by the neighborhood cats. The pathways are not visible like the trails of deer and the runways of mice. It is as if the cat had learned not to beat a path to the game and so betray its presence. The only way to find cat paths is to watch the cats walk along them. One trail in my yard threads beneath overhanging rhododendron, passes the edge of the flower garden, and arrives at the sandy area near the toolshed—a sunning spot. Another skirts the grape arbor and arrives in the small field beyond my woodpile, where mice abound. Only one crosses the open yard, and this would not be there were there a more secretive way to get to the bird feeder.

On their pathways cats avoid other cats at all costs. If by chance two arrive at a crossroads at the same time, a despicable situation, they prevent a confrontation by communicating with their eyes. No sound is uttered, nor are teeth bared. They stare, sending a message over as far as one hundred meters with their large eyes.

I saw the stare keep two toms from meeting while I was sitting with a friend on his roof patio in New York City. The two cats were walking across the rooftop on two different pathways destined to cross. One cat had a torn ear, the other a scarred cheek, attesting to their free-roaming life among scrapping city cats. They were on a collision course at a brisk pace. Anticipating a fight, I sat forward. Within two feet of the crossing both stopped, sat down, tucked their toes under their chests, and stared at each other.

After a long time, ten minutes at least, Torn Ear looked deliberately away—then back.

Several minutes later Scar Face looked away—then back.

*The staring match*

This went on for almost an hour, each staring and looking away, each feeling out the other and talking threat talk through the eyes. The tension was building. Torn Ear looked away for the umpteenth time, and Scar Face got to his feet. Hesitatingly at first, then swiftly, he ran down the path and dashed across the crossroads. He disappeared among the TV antennas surrounding a water tank. After sitting for another half hour, Torn Ear got up and followed the same path. Scar Face had won the battle of the eyes, but that was not the point. A confrontation had been averted.

**STRONG MESSAGE**

Even the discomfort of staring matches is often avoided by cats' form of scent marking. Cats spray urine to make their environment familiar and to leave messages for others.

Both sexes spray urine—the females spraying more rather than less than the males, but in smaller quantities. Since the structure of the female does not allow her to spray in jets, as does the male's anatomy, she simply emits droplets. Males spray by backing up to an object, lifting their tails high, and somehow turning their penises in such a way that the jet comes out parallel to the ground—and backwards! Tomcats, but not females, spray not only path and place within their second-order territory but also objects in their first-order home. Although full of subtlety and pleasantly odoriferous to the cat, cat urine simply stinks to us. The smell of urine from unaltered pets is particularly strong and hard to eradicate. Altered pets, which no longer produce the hormones that advertise their sexuality, have milder urine, and the males don't spray the house.

As for the meaning of spraying—a bit of spray exuded on strange territory fills the air with the reassuring scent of self and gives cats confidence to proceed. Their own spray evidently also gives cats great pleasure. Trinket would spray and then return sometime later to sniff and rub her head in the odor as if enjoying herself in the way a pretty woman admires her reflection in a mirror. At the same time, the odors are messages to other cats; they say that the marker is nearby or has passed that way. Apparently the intensity of the odor tells the reader how recently the marker passed by, and it can help the reader judge whether or not it is prudent to continue along that path or sit and wait for the scent marker to finish its business and leave the vicinity. Cats sitting for hours under a bush are not necessarily hunting. Very often they are sniffing fading pheromones and waiting for the air to clear, so to speak. The scent is certainly not a keep-off-the-property odor, as is the dog's or wolf's, for I have seen both males and females sniff a post that Trinket marked and sit down quite contentedly nearby, even when she was not in heat.

We do know that urine messages bring mates together, however. A female's spray is droplets full of news of her sexual readiness. She tells the males that estrus is beginning and gives them progress reports; they answer in kind. In addition, she states who she is, how old she is, and her mood. Cat urine varies in strength according to the emotion of the cat. An angry cat and a neglected cat emit strong urine whose odor is a concentrated essence of their mood.

Certain odors, usually sexual, when inhaled evoke a strange and sort of disgusted-looking grimace and noise called *flehmen,* a German word with no English counterpart. The cat sniffs, sometimes touches the scent substance with its tongue as well, then raises its head, pulls back its lips, wrinkles its nose, inhales deeply, and holds its breath. The strange sniff is a thorough examination of the material by both the taste buds and the nose; it is the quintessential sniff.

We probably miss most such expressions, since a cat is more likely to come upon an intensely interesting odor in its outdoor wanderings than in its first-order home, but occasionally a cat will *flehmen* if you have been with another cat. How your cat reacts to such news depends on whose odor you have brought home. Danny never failed to suck in, pull back his lips, wrinkle his nose, and gag when Joan brought home the odor of Jerry, an enemy. Danny was not so much angry at Joan for visiting his enemy as at the cat who was there in pheromones. Smell is reality to a cat.

House cats and many of their small wild cousins generally bury their feces. When they don't, leaving them instead out in the open, they are intended as communication. Apparently feces on display are simple statements, for while cats investigate them with some curiosity, they neither run away from nor actively seek out the sender. Perhaps they are "ain't-

I-somethin' " posters. Morris Hornocker, wildlife professor of the University of Idaho, reports that the Canada lynx and the puma not only deposit feces in the open for others of their kind to read but kick up snow and earth to set them upon—the better to display them. In Big Cypress Swamp, Florida, I found a bobcat's feces conspicuously deposited on a mound of once-beautiful royal ferns that the animal had kicked and trampled into a pile. I think I responded to the message just as another bobcat might have done. Wow! Look at that, thought I.

Feces left out for people to see can, as some of us know, be a blue-funk message. Danny left such a statement on Joan's favorite chair after she punished him for getting up on the table and eating the fish fillet. Trinket deposited the same remark on my daughter's bed when she brought home a stray cat.

If fecal messages can be resentful, scratch messages can be downright vindictive. To keep their front claws honed, cats scratch furniture and trees to remove the outer cover of their weapons, peeling them down to sharpness much as we sharpen pencils with a knife. They bite off the claw covers of their rear paws. Sharpening the claws, then, is a preparation for killing, and as such it becomes a declaration of war. Danny often sharpened his claws after Joan scolded him. He would run to the chair upholstered in needlepoint, reach as high as he could, twitch his tail, unsheathe his claws, and dig in. The sounds of threads breaking as he dragged the length of the chair arm were terrible to hear. At other times he claw sharpened to show Joan how strong and wonderful he was. He would wait until she was watching him, then his back would ripple and his eyes would soften; the meriow he gave as he dug in was the sound of the courting tom. She had answers for both scratched communications—a maga-

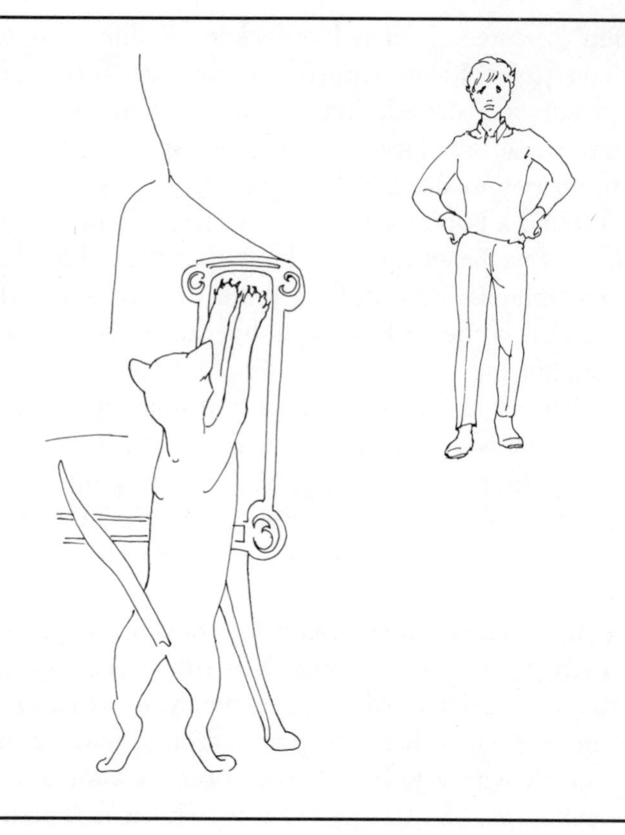

zine rattled and tossed in his direction.

Claw sharpening is carefully observed by other cats. The neighbor's black cat would sit under a bush and stare at Trinket when she tore grooves into the apple tree. When she was done, he would get up and walk away, with apparent lack of interest, only to mosey back later to read her grooves. No doubt the height of grooves gives an indication of the maker's size, and the depth a measure of its strength.

Occasionally the cat's hind feet get into a scratched communication in what is known as *paw wiping*. The cat vig-

orously scratches the ground with the hind paws. It is an act of extreme hostility because the back feet are used primarily for vaulting the animal forward into battle or tearing open the belly of an opponent. When they are used, it is serious talk—of a kind that some people can get into with some cats. Joe, a hired man on our Pennsylvania farm, kept the tomcat out of the kitchen by wiping his feet and standing broadside to him, as cats stand to others when paw wiping. The tom would take stock of the movement and, ears pressed to his head, slink off along the wall of the house, headed for the barn.

**TALK TO ME OF LOVE**

It is a seeming paradox that although cats despise each other and avoid contact whenever possible, they truly need our attention. Cats must get affection from the person or persons they live with, or they become depressed and will not groom themselves. A cat that is simply given food may live, but it is usually dirty and seedy looking.

The paradox dissolves on closer inspection. No matter how solitary the fashion in which an animal must earn its living as a hunter, it must be social enough to mate and to raise a family.

In the wild, female cats accompany their kittens during their first five months of life in order to adequately train them to hunt alone. Since they give birth every year, they are by themselves only a little more than half the time. There is also evidence that female relatives—mothers, daughters, and sisters—share a territory more closely than males. Witch, a "house" cat who scorned a roof over her head, produced annual litters simultaneously with her grown daughter. While her daughter nursed both her own and her mother's kittens, Witch hunted for the whole extended family. What's more, Paul Leyhausen noted that his females tended to mate with favorite males. Other observers, too, have thought that free-roaming females prefer to mate again and again with the same male if they are allowed the choice. Trinket's many litters looked a lot like Belknap, a gray-and-white tom in the neighborhood, and we assumed she was faithful to him. Although males do not help raise the young, some do pay calls on their family. Belknap turned up at the door with a chipmunk one day right after his kittens were born. Cats do love.

In fact, because we have to confine or alter males and females to control their prodigious reproductive rate, we may cut off natural social interludes in cats' otherwise solitary lives, leaving them lonely as well as loners.

Occasionally cats get together outdoors in an odd catlike manner that is perhaps an awkward attempt to socialize. They gather in considerable numbers, sitting well spaced with their feet tucked under them, smiling in some sort of cat communion. This is about as much of a concession as cats give to cat social events, and after a few hours they become bored, get up, and walk away. Such a gathering took place in a field near the station platform in White Plains, New York, at least once a week during a spring when I commuted early enough to see it. At about an hour after sunrise, I could see cats coming from all directions—out of the parking lot, down the street, meandering from the row houses, and strolling up from the streambed along the tracks. They were dignified, walked slowly, and used their eyes to settle spatial quarrels. Now and then one would spit to hold its place. They were always gone when I came back at the end of the day.

In your house you fill social needs by simply being there. Most times social contact is casual, but it can be passionate, as when Danny demanded the attention of Joan after her two-day absence

It is that passionate demand that is so startling in so aloof a creature. Geneticists suggest an explanation. The genetic change that relaxed the adult cat's wariness of humans affected, they believe, maturity in general. The young of most mammals are curiously unafraid and only playfully aggressive. That mildness is eventually overwhelmed by the more cautious and defensive behaviors acquired in adulthood. Domestic cats, it seems, do not develop fear and fury toward people. Without that the domestic cat reveals to us a kitten's affection for its mother, a mother's devotion to her young, and a lover's passion for its mate.

Most of a house cat's vocabulary of affection is derived from kitten expressions. By watching mother and young, the entire vocabulary can be seen and interpreted as she and her kittens, and they among themselves, interact, using the mews, poses, and actions that the grown cat later lavishes on its persons.

**BACK TO THE CRADLE**

The pregnant house cat is true to her wild heritage. She picks a den in her first-order home in which to give birth to her young. The den may be a closet, basket, box, or some nook among the ducts and rafters in the basement. Most females, having established a nest, tend to return to it year after year. Trinket's birthing den was the clothes closet in Twig's room. She demanded shirts, socks, and blue jeans for nesting material by scratching the floor of the closet and meowing loudly.

Some cats give birth so privately that the event is only discovered by happening upon the kittens. Others want at-

*Announcing the hour of birth*

tendance. Trinket announced the hour of birth to us by mewing, running toward the nest, coming back if we did not follow, and running forward again. She could hold back the birth for several hours while she gathered her human friends. Many of my friends report that they, too, are called to their house cat's birthing. As far as is known, no wild or feral cats behave that way. They give birth alone and in silence. Perhaps it is the kitten in domestic cats that wants attention from the mother in us at this time.

Immediately after a kitten is born, and sometimes while the hind end is still in the birth channel, the mother licks open its nostrils and cleans the mucus and blood from its fur.

Some people object to a cat's licking them, because of the roughness of the tongue. But licking speaks of the intimate relationship that originated between mother and kitten, and it is bestowed upon people who fill that special role of family in a cat's life. It is more than affection; it is bonding and should be endured with that in mind.

Kittens are born appealingly furry, but the eyes are closed as well as the ears, so communication must be largely by touch. We ourselves find them so adorable that we can hardly resist fondling and holding them, but they also speak warmly to one another. Before birth, the skin around the mother's mammary glands has lost its fur and become highly vascularized so that the nipple area is softly nude and very warm. When the mother enters the nest, her kittens crowd to her belly and, having located by touch and scent their preferred nipple (which they keep for the duration of nursing), knead at it with their forepaws to stimulate the let-down of milk. Leyhausen calls this the *milk tread.*

The kneading of your adult cat's paws on your lap or arm goes back to the relationship between mother and young and is a request for comfort.

Although usually a warm and friendly gesture, the milk tread can also be hurtful and demanding. Trinket on occasion at night while I slept would leap onto my stomach and begin to softly knead, demanding something—usually to go out. If I ignored her, the pressure became harder and harder, and the purr would shift into a purr-snarl. If I still did not get up, the kneading became so strong that it actually hurt, and I would be forced out of bed to open the window and let her onto her pathway that led across the roof and down the lilac bush.

For the first few days of life, kittens huddle together, one head upon the neck of another, in a compact circle. If the mother runs off and a kitten is still attached to her nipple, it may drop quite far from the nest with its circular snuggle of kittens. It finds its way back by an inherited, but seemingly intelligent, behavior. It mews pitifully, a high-pitched tone incessantly repeated. If this fails to bring the mother, the kitten begins a miraculous search for home. It crawls with shaky movements, not on a straight line, which might not be the right direction, but in a spiral. By spiraling in an ever-widening circle, it gets home eventually. A cat will also spiral your lap or a soft pillow as if looking for the center of the snuggle target.

For her part, a mother cat draws her kittens into a group with her paws. This all-in-the-family gesture is also practiced on humans. When a cat is picked up and hugged by a member of the household, it will pull in its claws, cup the paw, and extend the whole foot up and around the shoulder as if drawing the person into the circle as it was drawn into the kitten group by its mother. William, although a male, uses the same gesture to bring Sara's arm into the circle of the nest he occupies in her lap.

Less delightful is the bumping way of begging that kittens develop as they are weaned. Trinket's kittens begged for milk by running alongside her at full speed, bumping against her flanks as they traveled and knocking her off her course. Adult cats are whizzes at using this begging behavior on people, dashing along beside and bumping against us until we grant their request, trip over them, or toss them out the door.

## MEOWS AND PURRS

To the uninitiated, the cat's meow is a single all-purpose word that means everything from the food that cats "ask for by name" to a request for sexual attention. Actually, there are numerous meows that differ in pitch, rhythm, loudness, tone, and "pronunciation," and each has a different meaning or is spoken only under certain circumstances.

A mother cat, for instance, greets her litter with a chirruping sound whenever she returns to the nest. She does this from their earliest hours, when it is doubtful they can hear her through their still-sealed ears. That chirruped meow, rendered *mierrrow* with the *r*'s, burbled, is a great compliment when directed to a human, for it is an intensely affectionate and personal greeting. Uttered with a greater passion—and often so incessantly as to drive one to distraction—it is meant as a come-on. The cat wishes to mate. In other words, some degree of intimacy is always implied by the chirruped *mierrrow*.

A kitten's first sound is *mew*. It is weak, high-pitched, and brief during the first days, a complaint at being separated from the huddle and a plea for help. This mew remains as the polite form of request but also develops into the more demanding—or even commanding—meow. In its superpolite form, mew is actually soundless, a mere miming with no participation of the vocal cords. The mouth opens, the head goes back, all the gestures of mewing are there, but no sound comes out. Several cat owners interpret the silent plea as a smug way of demanding food at the table without being obnoxious; others find it to be part of the telepathy between a cat and its special person.

Trinket mimed soundlessly with mouth open to announce her presence among her kittens when she and they were at peace—which says something for the telepathy interpretation.

The paws also send messages through the floor or ground, for the cat does not always walk quietly as the fog on those "little cat feet." Trinket called her kittens by striking the floor when she felt a meow was too loud and dangerous for the occasion. She also used the floorboard telegraph when the house was noisy with talking people and her meow could not carry above the din. Apparently the vibrations could carry, and she would thump up the steps, sending signals through the wood to her kittens in the closet. The kittens would feel the message, awake, and run to meet her.

I eventually learned to call the kittens with a tap on the floor, but never Trinket. Apparently the adult cat outgrows its response to this command, although it still gives it. William, whose only experience of kittenhood was his own, thumps out a heavy message on his way downstairs to call his family to dinner. That is, they are to assemble to give *him* dinner. At twenty pounds the message is loud, yet the huge cat can pussyfoot the stairs when he wants to.

When vocalized, the meow comes off the forward part of the tongue—the same place where we shape our imitation meows. It is used all through life when asking for care, such as a bowl of food, or to be picked up and caressed, or for other instructional messages. The basic request is made while the cat looks at you; it is a repeated series of smooth whines that actually sound like begging.

But the expression varies considerably with the intensity of the cat's felt need. The loud, clipped meow is a demand: "Let me out of this house" or "There is no water in my bowl."

Cat lovers who really listen to their cats can distinguish a more extreme form of demand meow that might add to the request to be let out—the further distress note "I have to pee right now." Patricia Moyes, author of *How to Talk to Your Cat,* describes such a demand meow as "high-pitched, sustained, incessantly repeated and absolutely infuriating." Of course, that's why it works.

Moyes further distinguishes the complaint—"basically the high-pitched demand, but laced with a plaintive, falling cadence"; the protest—"the equivalent of a child's whining—a plaintive, high-to-low cadence, uttered without too much conviction and much lower in pitch than the complaint"; and indignation—"the chief characteristic [of which] is self-righteousness. It is a single high note uttered in short, sharp syllables and indicates that the cat has a legitimate grouse that he feels would stand up in a court of law."

Trinket had at least five different meows when she talked to her kittens. The two most understandable were loud—one that meant, "Follow me," and a sharp nasal one that meant, "Stop it." If I imitated her "Stop-it" command when she got up on the table, she would jump right down.

She used a high-pitched, open-mouthed meow to call the kittens to solid food, a sharp growl-mierow to send them to cover. I learned to imitate the growl-mierow, hoping to keep Trinket's kittens off the road, but I never had to use it. She had already communicated the hazards of roads and cars. They hid at the sound of a motor.

The whole series of meows, from timid request to arrogant command, is easier to understand than to describe. Most cat owners can tell by context what a cat wants and how badly, for it won't urgently request to be let out while

## MEOW CHART

*Kittens have only one word, but its intensity is a clue to the urgency of their need.*

| | |
|---|---|
| *mew* (high-pitched and thin) | A polite plea for help |
| *MEW!* (loud and frantic) | An urgent plea for help, as when the kitten is lost |

*Adult cats elaborate on the basic mew and make their meaning clearer by poses, gestures, and context.*

| | |
|---|---|
| *mew* | A plea for care and/or attention |
| *mew* (soundless) | The very civilized, polite form of the plea |
| *meow* | The somewhat more demanding, begging form of the plea |
| *MEOW!* | A command to give care and/or attention |
| *mee-o-ow* (with a falling cadence) | A protest equivalent to a child's whine when he has not gotten what he wanted |
| *MEE-o-ow* (whined, but shrill) | A stronger protest |

sitting next to the can opener, nor will it mew politely for food at the door. Still, for the most interested, a tape recorder is a good way to catch, play back, and memorize subtle differences. Some who have also recorded their cat's greeting to them have discerned that it can be the equivalent of a name. Just as you have a personal call for your cat—"William" or "Here, kitty"—your cat has a personal call for you. Taping it will help you to recognize it, for it is always the same unique meow.

When kittens' eyes and ears open at about five to ten

| | |
|---|---|
| *MYUP!* (short, sharp, single note) | Righteous indignation |
| *MEOW! MEOW! MEOW! MEOW!* | A panicky call for help, as when the cat is being molested |
| *mier-r-r-ow* (chirruped with a lilting cadence) | The "hello" greeting, implies intimacy when uttered with passion |
| *Tom cats have a vocabulary of their own.* | |
| *RR-YOWWWW-EEOW-RR-YOW-OW* | Night song when on the sexual prowl, known as the caterwaul |
| *merrrow* | Call to a young male to come out and fight |
| *meriow* | Courting call to female |
| *Mother cats have special calls used only with their kittens.* | |
| *MEE OW* | Call to come for food |
| *meOW* | Call to follow |
| *ME R-R-R-ROW* | Call to run to cover |
| *mer ROW!* | "No!" or "Stop it!" |
| *mreeeeep* (burbled) | The "Hello" greeting to kittens and the kittens' disarming greeting to adult cats |

days old, they begin to be able to cry in panic. Moyes describes the adult's panic cry as "a frantic squawk . . . [that is] blood-chilling and [that] must be taken seriously." The cat is in trouble—the neighborhood kids have tied a tin can to its tail. Kittens panic under less dire threats. A kitten calls frantically if it is lost, stuck behind the sofa, or being attacked or trampled by siblings. It is a long-distance cry. Trinket could hear it from upstairs when we brought a kitten downstairs to play with and admire—a circumstance she was likely to interpret as a kitten either lost or under attack. She would come to its "rescue" on the run, calling the growl-mierow "Run-to-cover" command, which, in this context, we took also to mean, "Give me back my baby."

The purr distinguishes cats of every kind from all other animals. It is unique in nature. Charles Darwin wrote that of all the feline sounds, "the purr of satisfaction, which is made both during inspiration and expiration, is one of the most curious." Lonely people feel wanted and good again when their cat purrs in their laps; it is a sound that needs no translation. The little girl across the street heard Trinket purr for the first time and told me my cat was smiling out loud. And so she was. Of the approximately sixteen different sounds the cat makes to express emotions and desires, the purr is the only one that conveys luscious contentment.

What is particularly wonderful about the purr is that it is given only in the presence of another animal. A lone cat sitting in the window does not purr, no matter how content. Cat purrs are communications, and cats are too sane to talk to themselves.

Purring is first heard when the kitten is about two days old and nursing. From that first contentment purr, the sound develops a more varied range of expression and meaning. A single purr announces that a pleasurable contact has been made—you have touched the cat's head or back for an instant—whereas a rhythmical purr repeated like an ABA song form (A: *rrrr,* B: [cranked a bit higher] *rrrrr,* A: *rrrr*) declares that the cat is being held and generously petted. When the cat is wonderfully content, it can purr loud enough to be heard across a room. But the classic ABA purr is only one of several murmur sounds that are made in the throat and that, in varying intensities, signify either lessening contentment or rising excitement. When an ordinary purr trips over into a purr-snarl, it says, "Enough of this sweetness. I'm annoyed."

Heightened excitement is expressed by the abrupt loud purr given by a female when she is ready to be inseminated. That word is not spoken to humans.

**BUZZ OFF, FELIX**

A mark of our distinction among cats is that they so often direct to us their positive statements and so seldom their negative ones. We expect dogs to resent our entry into their territory, but when was the last time a cat growled at you?

Cats are not so tolerant of one another.

It is not that cats cannot live together in a human household. They do, even purring in concert and grooming each other, but such cats are usually litter mates, or one of the pair was introduced as a kitten. Only on rare occasions do adults adjust to each other peaceably in the same household. Any cat given its druthers prefers to live alone with you.

To understand what would happen to your cat if it had to depend on another cat for sociability, you need only to observe an encounter between two strange cats.

Riley is large and yellow, and he is battered—a sign to other cats that he has survived many battles and has the nicks and scars to prove it. He met the staid, black Parson when two friends of mine together rented a mountain cabin for the summer.

On their arrival the carrying cases were opened at the same time. The cats ran out and faced each other across the room. As if by agreement they both turned away and inspected the new environment to establish their first-order home. But the question of who was boss soon overshadowed even that important duty. After a brief look at the surroundings they faced the real problem: each other.

First they sniffed nose to nose without touching—the

initial part of the cat greeting that every cat owner is familiar with. It is said to you when the cat jumps in your lap and lifts its head toward your nose, sometimes sweetly touching for a moment. If it is feeling particularly friendly, it will turn its rear end toward you for an anal sniff. This is an honor that most people shun.

Red Riley and black Parson were not going to honor each other with anal presentations. They stood apart, sniffing noses faster and faster. Angrily they closed their whiskers tightly against their faces. Both lowered their hindquarters a little bit and equally. Moments passed; they tilted their heads at an angle. They held their ears erect and faced each other, feet planted firmly. They stared the threat stare—pupils dilated, eyes wide, the lips lifted slightly. Each was trying to psych out the other, like wrestlers in the opening moments of a match.

Parson wore down first. He lowered his haunches ever so slightly. Had these two been wise cats, that would have ended the battle for dominance. Those lowered haunches foretold the outcome, for the gesture is made by the loser. I had seen Trinket use this code on me when I reprimanded her for being on the table. She was saying. "You win."

Parson's admission, however, did not end the conversation. Riley was aching for battle, and Parson responded. They moved in closer until each could sniff the nape of the other's neck, the prey's death spot that every cat knows so well. They felt each other's nape with their whiskers, as if through these—their stethoscopes—they could measure the impulses of nerves running along the spinal cord.

The sniffing of the napes moved to the flanks. Riley was in charge. He had not missed the lowering of Parson's rear end, and he was confident. He showed his confidence by lifting his tail and granting Parson an anal sniff.

The odor and the act of his self-assurance put Parson on the defensive. He flattened his ears, twisted them back and down, and crouched. Forced to look up at Riley, he was at a disadvantage and, now aroused to fear, acted to protect himself. He growled. Low-pitched at the outset, Parson's growl, like all cat growls, was a curious sustained noise coming from the back of the throat that rolled out louder and louder—a sound that falls somewhere between a dog's growl and a very deep meow.

Then Parson struck Riley on the nose. It seemed that he was the aggressor, so furiously did he strike, but that was not so. The paw strike is a defensive action, and it gave the black cat time to turn and run far enough away to feel somewhat more secure. He pivoted and faced his still-serene rival.

Riley walked to the spot where Parson had been sitting and sniffed. "I'm in charge," he was reiterating, as he lingered long over the odor, before slowly approaching Parson. Head high, tail twitching, Riley went through the greeting ceremony again, sniffing nose, flank, and anal area. This was an insult. The greeting was being used to rub in his superiority.

Parson felt the psychological bruise. He answered with the spit, that sharp hissing sound that we use ourselves when angry and frustrated. We also use the same facial expression as the spitting cat—lips curled up to expose the teeth, nose wrinkled, eyes slanted downward.

Having spit out his wrath, Parson turned and slunk off, and as he did so, spots on the backs of his ears, which some cats have, gleamed like eyes. They are called *threat spots* and can disconcert a rival. Riley saw them and hesitated just long enough for Parson to jump to the top of the couch.

His height reversed their roles; Parson now was, literally, in a position of dominance. He eyeballed Riley. Riley sat down and looked away.

When at last Parson forgot the feud and jumped down from the couch for dinner, Riley ran out from behind a door where he was waiting and watching, biffed the black cat across the face with a paw, and rolled beneath him.

He had gained the offensive position in a cat fight. Slashing his unsheathed claws, he raked the belly of wailing Parson and slashed his elbow with his teeth. When fighting each other, cats are programmed to cut the elbow with their teeth, not grab the death site at the nape of the neck. They will injure but not kill each other, except in rare instances of error. Bloodcurdling screams issued from both cats, and then they broke apart.

After a pause Riley attacked again. There was no breaking them up without my friends' getting slashed and bitten. Tufts of fur flew through the air, claws ripped in movements too swift to see. When they broke apart this time, Riley was clearly and permanently the dominant cat. Parson thereafter put his ears back and down in his presence.

They never fought again, but Riley would not let the subdued Parson forget who had won. If Parson walked within twenty feet of him, Riley gave him the threat stare and growled or spit. Parson would lower his head, neck, and ears and slink off. They never got along, but they had arrived at a working arrangement. Parson had to stay away.

Apparently cats never think of even bossy owners as rivals or enemies, for they reserve their growls, hisses, and spits for the vet who must do a painful examination or for a person who abuses them. Female cats do, however, speak to their owners quite passionately of sexual love.

### THE SEXY COME-ON

An unaltered female confined to the house without a partner flirts with her human friends, both male and female, as she comes into heat. She rubs her head on the floor in their direction and lies broadside to say, "I choose you as my love."

As the estrous period advances, the days pass in many such graceful mating overtures until she is ready to be inseminated. Then she announces her condition by lying on her belly, forelegs outstretched from the elbows, her neck extended above them. Her eyes narrow to a seductive slit. Her ears are folded daintily back. Her rump is raised, her knees are practically on the ground, her heels are up as she lifts her rear on tiptoes, ups her tail, and holds it to the side. Trinket would talk to my daughter in the mating language even when she wasn't in estrus by directing one or all of these movements toward her. It was a language we learned to read and which Twig found very flattering. She would answer Trinket's overtures with soft words and strong kneading down the back like the caresses of the tom.

You can elicit similarly sensual behavior in most cats with catnip, the one means by which humans can communicate with their pet through scent. Catnip, a weed to some and a valued herb or ornamental to others, contains a volatile

oil that, it is suspected, in some way resembles cat sex pheromones. The Japanese plant *matatabi,* the silvervine, and one of the garden heliotropes of Europe also contain volatiles that have an erotic effect on cats. Although some cats are not susceptible to these odors, most cats, of both sexes and regardless of whether or not they have been altered, go into an exotic performance, rolling and slithering as if courting. They claw, clutch, rub, and lick a toy stuffed with catnip.

John Henry, an Abyssinian kitten, went into similar conniptions over a perfumed guest. The perfume was Tabu.

Trinket was a beautiful dancer when she was given her catnip-filled mouse. She would rub her neck on the floor, slide along on her shoulder, then roll to her side with a thump, jump up, bat the toy, pounce, and go through the sensual

*"Be my love."*

routine again—very much like the routine she performed when she was in heat. Her reaction to the catnip, however, did not bring on her period of estrus, as some people might believe. When I took it away, she was no closer to mating than before the plant fatale was presented.

Alluring poses and gestures are only part of the mating vocabulary. We seldom get to see the rest of it, for cats usually mate at night. With persistence, however, I was able to take notes on one of Trinket's beautiful conversations with her mate, Belknap.

Toms have sexual periodicity, although not as definite or obvious as the females, so they are not always ready to mate. When this particular courtship began, Belknap, attracted by Trinket's estrous perfume, had arrived at our porch, but he was not very interested. Trinket took the initiative. She rubbed her neck on the floor, then she rubbed her shoulder, rolled onto her side with a thump, and faced her lover sideways. She had to repeat this seductive performance over and over for several days before Belknap began to show interest by spraying.

He brought Trinket a present, a mouse that she coyly hugged to her belly while lying on her back. She kicked the mouse with all four feet and batted at it with her forepaws. While she played, Belknap pounced on her. She was not ready. She swatted him and leaped away. They locked paws, rolled, got up, and chased each other up and down the front steps and around the bushes. They scrambled up tree trunks a ways and dropped back to earth. Now I understood that some of Trinket's games of chase with us were not just play but also the flirtation of courtship.

After days of chase and no action, Belknap became bored. He stalked off the porch to the garden. Trinket ran up to him and rubbed her head against his. That aroused his interest again. He attempted to mount her. She fended him off with a blow, screaming and shrieking as though he had attacked her. Belknap moved away, but not far. He sat under a small maple, keeping an eye on her. He mewed from time to time and, as males do during courtship to release tension, rattled his teeth together. He had not been home for six days. I brought them both inside to be fed. They were not interested.

Just before sunset Trinket gave the sharp purr-snarl that said she was ready. She looked demurely over her shoulder at Belknap. He stepped over her, placing his feet carefully alongside her body and gently taking hold of the nape of her neck with his teeth. Unlike the solid grip used in killing prey, he seemed to hold skin only. Despite the steadying grip, Trinket was not centered right beneath him. He stepped on her back and gave her the milk tread on the spine—the one Twig used. She righted herself squarely under him, and he, putting both hind feet on the ground, growled as he thrust. After a few thrusts he abruptly penetrated. She shrieked. The male penis is covered with horny spines, but it is not clear whether the female shrieks in pain or in ecstasy. At any rate, copulation is brief, and Belknap withdrew after only seconds.

His genetic code passed on, Belknap jumped off backwards and dashed away. Trinket turned and swatted at him, but he ignored her and sat quietly by. After a rest he took the initiative and began the long courtship again.

They copulated at least three times in the two days that I observed, but I could not be in attendance all the time; it was probably much more.

Whenever Belknap's interest flagged, Trinket, in spite of her apparent protest and pain, would motivate him anew by presenting her rear end, kneading the ground, looking at him over her shoulder, and even erecting hairs around the vulva to attract his attention and accentuate the target.

Finally, seven days after the courtship had begun and at the end of Trinket's forty-eight-hour ovulation period, the mating was completed and Belknap disappeared.

Based on the similarity of all Trinket's litters to Belknap, we believe that she was faithful to that single mate. But no one can say just how loyal or disloyal cats are, for they usually seek each other out in the darkness and cannot be observed. It is also known that if one male becomes exhausted during the breeding period or is dragged home and locked up, other toms will take over his job. It is this, more than anything else, that gives the female a reputation among people of being morally loose and the male the reputation of tomcatting.

Tomcats are, however, noisy in their ardor and fiercely competitive with one another. Just after sundown, when the mating cat's day begins, toms court fights as well as females. A coaxing *merrrow* given by boss toms lures young males out to fight.

Soon the caterwauling begins—that banshee call in the night that draws showers of curses and missiles from people whose sleep is disturbed. Even one tom caterwauling is enough to wake the dead. A group of rivals caterwauling is indescribable. Scientists think the call is to a female and yet concede that it does not seem to have an immediate effect upon her. Trinket never turned her head when the tomcats caterwauled, but hours later, before dawn, she would give the demand meow to say she wanted to go out.

### THE COMING-OUT PARTY

Even a rather aloof female house cat becomes unusually affectionate during pregnancy, as though preparing herself for the social demands of rearing kittens. Yet family life, that most sociable of times in a female's life, is brief with house cats.

The kittens, after a rollicking, tumbling kittenhood, are weaned with biffs and hisses and left to seek their solitary fortunes around three months of age.

There was a sharp contrast between Trinket's fiercely protective behavior when we brought a kitten downstairs to admire before she felt it was ready to leave the nest, and her equally fierce rejection of her kittens at weaning.

Trinket's rescue of her kittens began with a leap onto our laps. She would then weave her head back and forth as she honed in with eyes, nose, and ears on her young. With care she would grasp it by the nape of the neck, taking several test holds as if making certain she did not have the fatal grip of the cat on its prey. That the cat picks up its kittens with the same grip with which it makes a kill is worrisome until you realize that there is an automatic protection device. Nerves to the jaw pass on the "kitten" message, and the jaws lock apart. The murderous canine teeth are unable even to penetrate the skin.

After obtaining the right hold, Trinket would swing to the floor and, as she rushed up the stairs, beat out the foot tattoo to the other kittens: "I am coming."

Cats move their kittens frequently, not only for the purpose of good sanitation but also for the purpose of entering them into a broader environment for learning. The dark nest in the closet was good enough for wobbly offspring with thin mews and almost inaudible purrs, but when the kittens could meow for attention and had begun to play, Trinket carried them into the upstairs hallway and deposited them in the middle of the rug. Here they interacted with us, but only occasionally, because most of our family action was downstairs. The meetings were kindergarten lessons about people.

Gradually, as they became confident, she moved the kittens closer and closer to the main activity of the house.

When the kittens were something over a month old, she

carried them into the center of our life and deposited them in the middle of the living room floor. Here they learned the meaning of their mother's "No," the sharp *merROW*. It is clearly a no and possibly the reason that the domestic cat responds to our similar sound. Not only do theirs and ours both occur as sharp, short noises, but our faces have the same expressions when the admonishment is given. The forehead wrinkles, the mouth opens, and the eyebrows draw together. Trinket said no to such mischief as approaching the dog too closely, straying too far from the group, or disappearing altogether behind the couch.

The center of the living room floor was also the spot for the cat debut, a bittersweet ceremony held during the weaning period when the family is being prepared to break up and become loners. Swiss ethologist Rudolf Schenkel observed the debut among free-ranging lions in Africa. When the lion cubs are between three and four months old, the mother leads them from their nest in a dense thicket and across the savanna to meet the pride. As they cross the grassland, the adult members of the pride—sisters to the mother, young males, the father of the cubs—arise and wait quietly. The mother leads the cubs into the center of the gathering. The adults examine them with their eyes, ears, and noses, and occasionally with a gentle paw. Through all their senses they learn about the newest members of the pride—their sex, strength, and personalities. When satisfied they go back to what they were doing, which, with lions, is usually sleeping. Whereas before the debut the adult members of the pride usually have been bickering among themselves, all fighting ceases once they meet the cubs. It is much like the silence that ensues when a child enters a family argument.

The debut of the domestic cat is very similar. Only the pride is different; it is people.

On the day of the event Trinket would let us know it

was about to begin by wandering among us meowing softly, but demandingly. When she had gotten our attention, she would run upstairs, call, "Follow me," to her kittens, and parade them down the steps and into the center of the living room rug. There she would crouch with her feet under her. The kittens would look up at us and give us the silent meow. The first three or four times, we were not aware of what was going on, but because the kittens were so cute, we instinctively did the right thing. We picked them up and admired them. When I finally discovered Schenkel's research, many generations of kittens were off on their own with our blessings, even though we had not known we had given them.

With our new knowledge the last several litters were given grand debuts. All three children and I held the kittens,

*The debut*

sniffed their noses, and rubbed their heads. We stroked them on the tail, flank, and behind the ears to say in their own language that we were pleased with them. We also voiced our approval with high-pitched baby talk—sounds very close to cat sounds. My son Craig went one better. He lay on the rug sideways to them with his arm extended like a courting cat. There was no evidence that they were moved by this talk.

When we were done, or when Trinket thought we were, she led the kittens upstairs. Whereas before the debut she had conducted weaning mostly by noncooperation, nursing less frequently and more briefly than the kittens would have liked, she now said no to the first one that tried to nurse. Her irritation increased with each effort. *MYUP!*, she would complain when sick of saying the no *merROW*. If a kitten still persisted, she put an end to all demands with a spit or a growl. After the debut she paid less and less attention to them, and they wandered the house at will, filling the rooms with the wonderful sounds and sights of kittens.

Trinket's job was not yet over, however. All wild cats and many house cats train their kittens to hunt before bidding them a final good-bye. Trinket began by bringing a dead mouse to the nest when the kittens were about six weeks old, just before their weaning was begun. She would pick it up and drop it several times, growl over it, and then eat it, making no attempt to share with the kittens. A week later the kittens would sniff the mouse, drag it around, suck it, and also growl over it. At this point Trinket would bring a live mouse to the nest. She would now intersperse growls with soft coaxing purrs that stimulated the kittens to come forward and sniff. If she released the mouse, she would quickly recapture it as if to show them what to do. But, curiously, kittens do not take a mouse, alive or dead, by the nape of the neck at first. Like puppies that fight to establish their rank in the litter, kittens brawl during this period. Leyhausen thinks the killing bite does not develop until the kittens have established their pecking order and, with the fighting over, are no longer in danger of killing each other.

By three months, when the kittens were able to kill a mouse by themselves, Trinket was ready to let them go on their own into the world. In the wild, kittens would stay with their mother for months longer, following her along her hunting paths and honing by experience their innate hunting behaviors. That kittens are geared for months of practice is evident when you bring a kitten home, for much of kittenish behavior is actually hunting rehearsals.

## PLAYING CAT AND MOUSE

The house cat is an incredibly gifted hunter. It rarely misses, for its technique is not chase and overcome like the canids, but lie in wait, stalk, and then pounce only when there is no other outcome for the prey but death. Much of a kitten's play with us is a game of hunt in which we act both as the mother providing practice prey and as the prey itself.

Many cats do not outgrow these games. All the games Trinket played with vigor and exuberance as a kitten were still exciting to her as an old cat. That she continued to respond with kittenish enthusiasm convinced me that the geneticists are correct; the house cat is forever retarded in youth.

Cats are not the devastating bird killers they are reputed to be, simply because birds are too hard to catch. Cats get discouraged when they fly off, and soon give up. A Polish scientist analyzed the stomach contents of 500 feral cats killled by hunters. His findings exonerated the cat from the charge of being a ferocious bird killer. Seventy-four percent of the stomach contents was small rodents, 3.4 percent young rabbits, 19 percent kitchen scraps, and 3.6 percent carrion. An American study supports these findings. Of the 4,771 domestic cat feces examined on a California nature preserve, 21 contained the remains of rabbits; 8, the remains of birds; and 4,742, the remains of rodents—mostly mice.

The "mouse" you offer your cat may be a piece of aluminum foil on a string. Here are some notes on Joan playing mouse with Danny.

Joan rustled the "mouse." Danny sighted it, crouched, and ran speedily toward it in the stalking run. His body was almost flat against the ground; his forelegs made a sharp angle at the elbows and projected above the shoulder blades. His tail was stretched straight out behind, its tip twitching gently. His head was forward, whiskers spread wide, and ears erect and aimed at the toy prey. This is the run-stalk.

The prey lay still. Danny moved forward. Joan jiggled it. He prepared for a pounce by gradually lifting his rear heels off the ground until he was on his toes, his hind paws moving rhythmically up and down, rocking his whole hindquarters. His stretched tail twitched. He sprang, going for the mark not in one leap but in several bounds, keeping low. He struck the prey at ground level—the pounce.

The aluminum foil escaped Danny, and he positioned himself steadily for another pounce by grounding his hind feet securely. Stabilized, he was ready to follow any evasive movement. But now something happened that could not happen in nature. The "mouse" was transformed into a "bird" as Joan lifted the foil into the air.

Cats hunt birds by bending the paw at the wrist and sweeping the arm upwards. If you are sitting on the couch and the cat is under it, it will use this technique to catch you—and your attention. Had there been a way for Joan to dangle the toy below Danny, she would have transformed it into a "fish." The fish-catching paw and arm is a downward scoop with the paw bent like hook and the claws extended. Cats that sit on refrigerators and tops of bookcases "fish" for you when you pass by. I respond to all these overly realistic hunting games with a spit-growl. Sometimes it works, but not always. Cats love hunting games and practically nothing discourages them.

Sometimes more serious motives overwhelm the let's-pretend quality. My friend Gretchen Chinnock learned that from a tabby cat named Scruffy. He called on her every morning, asked to come in through the window, and slept on the couch. But after her daughter visited with her cat, Scruffy sniffed the room, the closet, and "his" couch, then stared at Gretchen and slipped under the chair where she was accustomed to read the morning paper. An hour or so later, having forgotten all about him, she sat down for her morning read. A paw shot out and up from under the chair and nabbed her, bird style.

As Danny's play bird rose into the air, he swatted, scooping upwards with unsheathed claws and spread toes. One claw snagged the foil. He held it, foreleg extended. Were it a bird, it would not get away.

You would think that when a cat discovered that this entrancingly moving object was only string and foil, devoid of warmth, fur, feathers, or scents, it would lose interest. But Danny, like many other cats, went on with the hunt, for he was as stimulated by the game as he would have been on a real hunt. He was exuberant. He flopped to his side, and his right foreleg seemed to come out of his shoulder as he reached for the prey. This movement said the aluminum foil was a mouse. Joan responded by making it act like one, keeping it as still as a mouse when a cat is around. Danny reached with the mouse scoop he uses to catch mice. Then, stretching his

foreleg from the shoulder, he lifted his paw, and prepared to smash down on the victim from above.

Joan gave the prey a hard jerk, and it flew past Danny's head and got away like a bird. Danny looked dumbfounded. His wide eyes and open mouth said so. Nothing is supposed to get away from *him*, despite the fact that it often does in games. His genes were stronger than his memory. Cats expect to be successful and are incredulous when they are not. The prey returned, moving like a mouse. Danny pounced and bit. That successful denouement ended the game.

A variety of cat toys and play techniques will evoke a variety of cat responses, adding to your and your cat's pleasure in games. You can drum up a search by rustling paper and scratching the floor with your fingernails. I squeak like a mouse by kissing the back of my hand and squeezing air through my teeth. Trinket, upon hearing these sounds, would come on the search from the upstairs bedroom down to the kitchen and meow in protest when she found only me.

A fake fur-covered or plush ball evokes the killing bite, given from the top, as though to the nape of the neck. The toy should be no bigger than about two inches in diameter. Domestic cats can catch any live animal not bigger than themselves, but rarely take on anything larger than a rat or pheasant. Trinket's favorite hunting toys were about the size of a mouse. We also liked to see her strategy for smaller prey, and for this we used a "butterfly"—a small piece of paper attached to a string.

Cats hunt insects, frogs, and toads, not with one paw as with a mouse, but two. When we fluttered the butterfly

toy above Trinket, she would hop, clap her paws together, then take a calculating look, leap, and close. This was not a favorite game, because it is difficult for cats to jump up and catch flying objects.

Rarely did she play rat hunt with me, a game that is the same as mouse hunt but with a larger object. The size of the "rat" turned her off. Cats do hunt rats, but with some reluctance, since these animals are vicious fighters and often turn and attack. For a cat to be attacked by the prey is not in its game plan at all, and when this happens, it goes into the anxiety dance—leaping backward, hind legs stretched stiff—then bolts.

However, you can get the flavor of how a rat is killed by using a smallish stuffed animal (one that is no longer wanted) on a string. Just as a cat attacks a rival by rolling onto its back, holding on with its front claws and slashing upward with rear claws, so it attacks a rat by rolling under it and batting it so rapidly and incessantly that it exhausts the beast. From the relatively safe position below the belly, it kills the rat with a bite to the chest or throat. Jerking the "rat" after your cat has a good grip on it will stimulate the belly-raking and throat-biting responses. It is not advisable to let small children try this, because if they get their hands in the way of the action, they will be scratched and bitten too.

Trinket was a ratter when the real thing invaded the house. She worked from nine o'clock one evening until dawn the next morning to catch a rat. Stalking, waiting, and finally driving it into the exposed springs of a swivel desk chair, she held it as if by a paw by never taking her eyes off the beast for one instant. At 5:00 A.M. the rat moved out, so still was Trinket. That was it. She grabbed it, rolled under it, and quickly killed the exhausted animal. She walked off and never came back to eat it. I tossed it in the garbage.

**THE REAL THING**

Several days after I took notes on Danny's play hunting, he found a live mouse.

For an hour he lay in wait for it, stalked, lay in wait, and then cornered it under the china closet, holding it motionless with his eyes. By not moving, a mouse vanishes from trouble. A cat needs motion to inspire a strike. All predators see motionless things, but not as well as moving ones. It is like someone pointing out a motionless deer in the woods to you. You barely see it, if at all, until it leaps off. Then it not only becomes clear but inspires a reaction. Some shoot, others pause and smile in admiration and awe.

Joan rattled the china closet. The mouse darted out. Danny saw it and pursued, bounding low to the ground. He pounced, slapped it down with a paw, and grabbed it by the nape of the neck—the cat's classic mouse catch. He bit. The bite was fatal.

Dead mouse before him, Danny did something that has puzzled cat watchers for ages. He laid the little corpse down on the floor, walked away, sat down some feet away from it, and proceeded to groom himself. Leyhausen calls this "taking a walk" and believes it is not so much a blasé attitude toward killing as it is a working off of tension. Little John Henry worked off tension under a different stressful circumstance. Attempting to defuse the aggressiveness of the many-times-larger cat to whose home he had been introduced, he would bat at his housemate while burbling fond greetings. Then, rebuffed by the old cat's baleful stare, John Henry would withdraw his paw and groom it avidly.

More puzzling still, and deeply upsetting to many, is the behavior of a cat that does not immediately kill its prey but only stuns or wounds it with a blow of the claws. William walks away from his stunned prey, returns, bats at it and tosses it as though it were nothing more than a toy on a string, leaves again, and repeats the cruel game until, probably by accident, the creature dies. Michael Fox explains this macabre behavior simply: It is a short circuit. Each step in the hunting sequence is triggered in the cat's brain by a stimulus. The stimulus to stalk and pounce is the prey's movement; the stimulus to give the killing bite is hunger. William is, by any standards, a fat cat. Moving mice trigger him to stalk and pounce, but his stomach doesn't trigger him to kill. Relieving tension, he takes a walk, sits down, grooms—and notices the moving mouse. Again he stalks, cuts off the animal's escape by those alternating paw swipes that toss it from side to side,

pounces—and leaves. His normal sequence is short-circuited by his full belly.

Danny tosses even a dead mouse, not because it twitches of its own accord but because, returning, his own pawing animates the corpse and triggers his swiping reflex.

Very often during the walk, prey that has been merely stunned escapes. Trinket invariably caught chipmunks by one claw, which did not kill but sent them into motionless shock. Presuming them dead, she would lay them down and take a walk, only to return and find they had revived and run away to see another day. It will come as a relief to those who can't bear to see suffering to know that a cat's attack releases in its prey a neurotransmitter that blocks pain and anxiety. The chemical, similar to morphine, renders the victim insensible to its predicament. It is alive only in a physiological, not a psychological, sense.

When Danny was stalking the mouse, his whiskers were fully stretched in the direction of the action. When, however, he picked up the dead prey, his whiskers enveloped it as if to protect it from thieves. The whiskers of a cat are organs of touch that bring in spatial information about the close-in environment, such as the size of holes or passageways through brush.

By touching cats' whiskers, and also what are called *sinus hairs* on the upper and lower edges of the lips, scientists have discovered a number of efficient reflexes that aid in hunting. Depending on which whisker, sinus hair, or lip area is touched, a cat will blink, turn the head away, or slam the jaws shut.

I picked up Danny's deserted mouse and examined it. The stab wound was miraculously clean. Only a small drop of blood revealed where this incredible surgeon had severed the spinal cord with a slash of a single canine tooth. Research has shown that the canine teeth of the cat have an automatic pilot. They hit the nape, slide down between two vertebrae, and, using the space within the joint as a guide, glide down to the spinal cord. Once there, nerves at the root of the canine tooth send messages to the jaw muscles. They close with enormous strength and with lightning speed.

I put the mouse down, and Danny clutched it. He carried it swinging between his two front feet, and he himself reeled slightly from side to side to swing it and keep it from hitting his legs. When he had transported it to the threshold of Joan's home office, he dropped his burden. That is as far as the prey of some cats ever goes. William the Fat brings the trophies of his hunts to his first-order home as offerings to the family, much as Belknap brought mouse presents to Trinket and her kittens, but he seldom dines on anything more hearty than the delicacies, such as mouse liver.

Danny plucked some of the mouse's fur, a final nicety before the feast, and licked the mouse's flank to clean off the plucked hair from his rough tongue. Gathering all four feet under himself, he crouched, the favorite eating pose of the domestic cat. He adjusted the mouse in his mouth, and then, using the carnassial teeth in the rear of his mouth as scissors, he cut the prey into small pieces. He swallowed them whole, neatly and without chewing. His paws did not enter into the action at all but remained tucked tidily under him. When the mouse was consumed—ears, toes, tail, and all—he licked his paw, and with it he washed his face, whiskers, and the backs of his ears. Gracefully he arose and walked off to his sunning spot on the windowsill. Slowly he closed his eyes, a smile on his lips.

## EXCLAMATION POINTS

Whether active or in repose, cat expressions are emphasized by their eyes, ears, tail, and even their markings. These exclamation points are worth careful observation if you wish to catch the finer meanings in cat expression.

The coloring of the face, tail, back, and belly of the cat emphasizes what is being said. White teeth are outlined by dark lips, and wrinkled noses are seen more clearly by the gathering of dark hairs in the grooves made by contracting facial muscles. The light belly, like the dog's, speaks of submissiveness when exposed, not to a leader but to a mate or rival. Some cats, in addition, have those eyes, or threat spots, on the backs of the ears with which Parson held off Riley long enough to gain the top of the couch. Cat expert George Schaller suggests that the spots are also "Follow-me" signals that shine out like eyes on a mother cat and guide kittens through tall grass and shadowy forest. Trinket, who had these spots, seemed to use them that way. I saw her take her kittens into the tall grass by the woodpile. They would deviate from her footsteps, get lost, leap up, possibly see those eyes in that moment of aerial suspension, and catch up with her. But perhaps they were zeroing in on her upraised tail.

The splendid cat tail is a semaphore that sends messages of friendship and warning, as well as "state-of-the cat" announcements. On windowsills and beside hearths sit cats with tails curled around their bodies announcing their comfort to everyone who cares to hear. When Trinket was walking about relaxed, she carried her tail loosely down, slightly out from her body, the tip curled gently upward—like all cats at ease.

After a slight rise in emotion, caused perhaps by a loved person moving around in the first-order home, the cat will express its cheerfulness by raising the tail, softly curved.

The greeting-tail hello with which Danny speaks to Joan is straight up and not curved at all—the flagpole. He greets me, however, with a slight reservation, holding his tail like an upside-down J. That crick on the end means, "I am not sure of you." Most cats give the J signal to strange people.

They feel friendly toward them, but not straight-up, all-out friendly. Reply to this by holding your hand out for a bout of sniff reading by the cat so that it can learn about your attitude and personality.

Danny tail talked when he had the mouse cornered under the china closet. The tip twitched and quivered to say, "This is my mouse. All other cats and people stand back." Beware that twitching tip; it is a warning. William tried to tell Sara's husband, Marty, not to bother him one evening. The cat was lying before him on the dining room table, twitching his tail. Marty reached out to pet William just as Sara saw the tail and warned him not to touch. Too late, the paw followed the "Don't-bother-me" warning.

While walking, a cat expresses this same message with a crick that begins at the base of the tail and moves upwards to the top, leaving a wake of bristling fur behind. As it reaches the top, every hair is standing out like a bottlebrush to accentuate the "I'm-going-to-swat-you" message.

That bristle tail is also a vivid signal in the Halloween-cat pose, a posture so magnificent that it has become the symbol of courage and defiance. It is best seen when the cat faces a strange dog. It stretches itself up to the full height of the legs and arches the back. The body hairs rise, and at the base of the tail, that crick begins, moves upward, and disappears, leaving the bristling flagpole. This is all offensive talk. If the dog is provoked to come on, defensive pronouncements, all in the same pose, are added. The ears flatten to the head, the mouth opens and exposes the sharp canine teeth, and the head draws in to protect the nape of the neck. The Halloween cat hisses and stands broadside to the enemy to appear bigger and more formidable. "I am ready to kill. I am ready to swat and flee," says this, the most expressive of all animal postures.

Tail emphases are mirrored by facial expressions as a

| THE CAT'S TAIL POSITIONS | | |
|---|---|---|
| *"All's well."* | *"I like you, with reservations."* | *"I like you all out!"* |
| *"You make me worried."* | *"I'm on the defensive."* | *"You're boss."* |
| *"I'm a humble cat."* | *"That mouse is mine."* | *"Got it!"* |

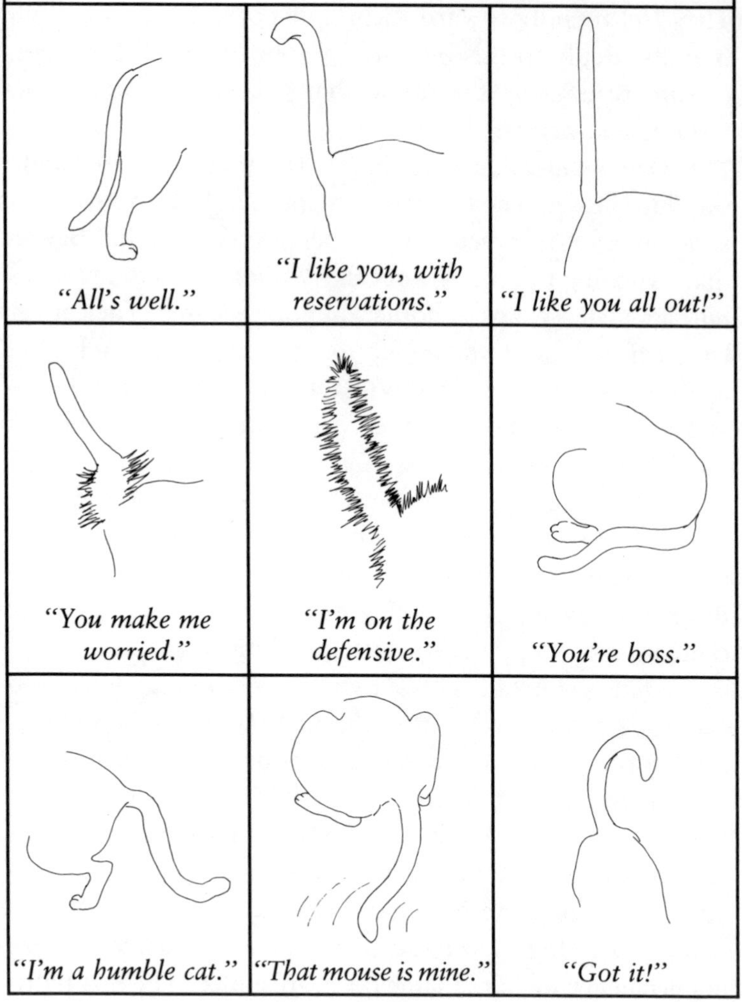

cat shifts from a pleasant mood to an angry one.

The cat face has six readable expressions. Each expression is an honest announcement of what is to come. A good cat-face reader can prepare for a swat, a bite, or a purr by quickly interpreting the facial announcements.

When your cat is simply saying, "I feel affection and want to be caressed," the ears are up and forward, and the head is tilted slightly downward. The pupils are almost round in the center of the eye. The lower lid rises slightly, softening the whole expression. The mouth seems to smile. The cat is prepared to purr if you stroke it or pick it up and hug it.

"Someone is coming," says the cat face with the ears up, eyes wide, pupils narrow, lips and whiskers relaxed. Danny would wear this face when another cat was approaching. As his concern increased, his ears would twist back until their openings faced to the side, and his pupils would dilate.

Unsure of who this "someone" was, he would shift to the offensive-defensive face. His head would lower as he stared. His ears would twist forward from their very bases, as if to pick up even the rustle of a whisker. His pupils would dilate wider. The meaning is ambiguous: "I might run. I might attack."

Step out of reach and watch for any further deterioration of mood once a cat states its uncertainty, for if it shifts toward the defensive-threat face, the cat is warning you of a swat and perhaps a bite that will draw blood.

As Danny progresses toward a mean mood, his ears flatten, his pupils dilate yet further, and his mouth opens, momentarily displaying his lower canines. Then the mouth closes, and the ears rotate forward but remain flattened. Finally, the defensive-threat face: Ears are rotated back now and so flattened that they are not visible from the front, the mouth is wide open, showing all four canines and the tongue, and Danny is spitting. His eyes become black as the pupils dilate completely. In a moment he will make good his threat to attack.

The cat's eyes, so expressively large for its head, are indices of its feelings. Research at the University of Chicago psychology department indicates that cat pupils, those vertical slits that expand in darkness and contract in light, also respond to the thoughts and feelings of the cat. The same is true of humans. Tests reveal that the pupils of our eyes dilate when gazing at a lover, when about to fight, and even when, of several available beverages, we sip the one we prefer. Our dilations are not, however, dramatically large. A cat's pupils dilate four or five times their former size when it is hungry and sees its well-supplied feeding dish. They also dilate when it sees photographs of a strange cat or of its favorite people.

We tend to think that animals cannot recognize objects in photographs, but a cat-sitter for a tabby named Rachel, owned by two Yale professors, told them when they returned from vacation that Rachel had sat on the dressing table staring at the family photograph for hours each day.

Cat eyes shine at night, throwing off colors that vary from yellow-green or blue-green to red. In the Middle Ages witch-hunters believed that the shine came from little windows that opened on the fires of hell, an apt description of the eerie flashing lights in the cat's eyes when the beam from headlight or flashlight strikes them in the dark.

The shine is actually a reflection from a mirror, a layer of cells of crystalline substance in the cat's eye. With this arrangement an image is "twice seen," once as light hits the retina on its way in, and again as it is bounced back through the retina from the reflective layer behind it. The double exposure helps the cat make the best possible use of available light under low-light conditions, such as when it is hunting at night.

That aloof look of the cat has a scientific explanation. We bring objects into sharp focus on a cluster of cells in the center of our eye called the *fovea.* Cats do not have this central point. They see primarily out of the periphery, which requires no focusing of the pupils. The impression given is of looking without looking, or of dreaming of visions. Many people find the far-off gaze cold and disturbing.

When a cat does stare at something for a long time, it either fears it or has designs on it. With the stare it is taking aim for the death bite. That is why cats themselves do not like to be stared at. Paul Leyhausen noted that when one of his cats was creeping up on its prey and realized another cat was watching, it instantly straightened up and lost interest in the pursuit.

Cats stare at prey and rival cats, but they do not look long into the eyes of lovers and mates as we do. Trinket never once looked for more than a second or two at Belknap, keeping those long looks for mice and rats.

She did not like her kittens to stare at her either. If one kept its eyes on her while she was coming toward it, she stopped, sat down, and looked away. Lion tamers, however, make use of the direct stare to control their animals. According to Paul Leyhausen, it works because the trainer's stare maintains his superior rank.

You, too, can talk to your cat with your eyes, but beware. It is not at all like looking into the eyes of your dog to

## THE CAT'S ANGRY FACES

*During an aggressive encounter the pupils of the eye, the angle of the ears, and the raising of the fur are clues to what the cat is feeling.*

convey love and affection. I kept trying to make friends with Freud, the psychiatrist's cat. One evening I was observing him as he sat atop the couch. My look, it seemed to me, was barely a stare, but before I could say, "Nice kitty," he was off the couch and speeding toward me like a launched jet. He clamped his teeth on my wrist.

On the other hand, a cat feels free to stare at you, and when it does, it usually wants something—food, the chair you are sitting on, your attention and affection.

Unbeknownst to many, people are talking to their cat when they read or write. By staring at a book or letter for a long time, we are announcing to the cat that we have something very interesting lined up. To a cat's way of thinking, we are inviting it to come and sit upon this fascinating "place" within its rightful environment. The cat obliges and sits down, only to be confused when it is picked up and dropped to the floor.

A better way to get rid of the cat on your paperwork is to tell it to sit somewhere else by staring at another book or piece of paper. Of course, you don't get much accomplished this way, for cats take their time in moving to this marching order. Furthermore, when you go back to your work or resume reading your book, you are telling the cat to come back again. It usually obeys. This can go on for hours. Cats have lots of time.

My friend Peg Bradfield, an illustrator, solved this problem once and for all. She had a tom named Jenkins who sat on every sketch as she worked on it. She tried to tell him to go elsewhere by staring at pillows and corners of her studio. He heeded, but only briefly.

One night she was working late to make a morning deadline when Jenkins, at her invitation in his view, jumped onto her drawing board. She did not have time to stare at another piece of paper, and desperation drove her to invention. She went into the kitchen, gathered every brown paper bag she could find, buffeted them open, and set them all over her studio floor. Cats *have* to explore caves. In the human environment caves are usually boxes and paper bags, and for many hours Jenkins went in and out of the caverns seeking whatever cats seek in such places. Peg worked peacefully, and just before sunup she put the last brushstroke on her illustration and arose. Jenkins was tumbling inside the last bag; but Peg's work was done.

Paper bags are the ultimate in polite, genteel communication with a cat. You can hiss, meow, even stare, arch your back, show your teeth, to no avail. But give your cat a paper bag and he bows to your wishes. In he goes—and gets lost.

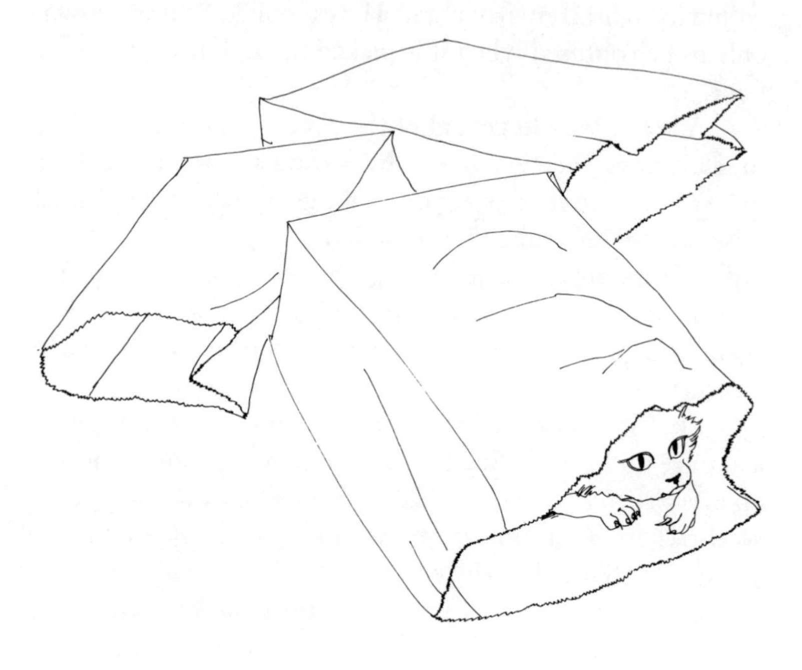

# The Birds

The television repairman pulled the set into the middle of the living room floor and leaned over to study the failed device. It was a routine job, and he was feeling relaxed and confident. Suddenly there was a whirr and something landed on his head. Not knowing what had hit him or why, he flailed his arms and pulled in his neck. A bobwhite quail flew off his head and landed at his feet. The bird's feathers were puffed, making him look nearly twice his size and a good deal more formidable than any four-inch-high bird should be.

*"Squawk, squawk, barker, barker, barker!"* the petite brown, black, and white bird screamed over and over. By the time Jean Bixler had heard the commotion and rushed downstairs, her pet bird and the man were facing each other in troubled confrontation. Cracker was screaming in earsplitting decibels, and the man was standing before him, his eyes wide

and his mouth open, signifying alarm in any language. Jean picked up the knight-errant.

"Meet Cracker," she said to the disconcerted repairman, then paused and smiled. "He's telling you to get off his property."

"That little fellow's talking to me?" The repairman grinned. "He's telling me off? A big guy like me?" He chuckled thoughtfully. As Jean stroked Cracker's black cap, the white feathers under his chin and above his eyes relaxed; the pupils that had dilated in distress returned to normal.

*"Get off my property!"*

*"Chip, chip, chip,"* she whispered to the bird. The man looked bewildered.

"I'm telling Cracker in quail talk that everything's all right." The repairman stared at the young woman who could talk to a bird and at the bird who could defend a house against a man. He returned to his work shaking his head in the silent communication that says, "I've seen everything."

## WHO IS MOTHER?

The secret to this unusual relationship between a person and a bird is *imprinting,* a form of attachment that can be elicited only early in a bird's life and that makes the difference between a Cracker and a wild, easily frightened bird.

Cracker was one of five eggs in a plastic incubator ordered by Jean Bixler from a Georgia quail farm. Twenty-three days after arriving, he was the only one to cut through his shell with his sharp egg tooth and survive. Wet and exhausted, he rested on his belly in the warm palm of Jean's hand until his feathers dried. Moments later he looked up with a penetrating glance and saw an object that would affect him for the rest of his life—Jean.

During that crucial stare Jean was imprinted on his brain as his mother, the living thing was was caring for him, keeping him warm, providing food and shelter. Also programmed into his brain cells in that same moment was the ineradicable vision that he looked like her and she like him. He was, according to the inherited wisdom of the birds, kin of the nurturer—in Cracker's case, a member of the human race. After being imprinted, Cracker was a person in his own eyes.

That is how it is in the eyes of birds. The creature that tends and nurtures it after hatching is "parent" whether it is a human being or a box of food pulled along on a string, as it was in an experiment with some goslings. The small brains of birds are programmed to seek the specific information in the environment that will help them to survive, and Cracker stored Jean's face as his first urgent point of orientation—"mother."

Person-imprinted birds make the most communicative pets, not because their bird vocabulary is different from their correctly imprinted wild counterparts, but because they assume you, as a member of the same species, understand them. They are naturally more willing to talk.

Cracker was imprinted strongly. That is typical of what are called *precocial* birds, those whose eyes are open at hatching, that are covered with down, are capable of locomotion on land or water, and can feed themselves. Such species—which include geese, chickens, ducks, pheasants, and peacocks, as well as quails—are mostly ground nesters. Upon stepping out of the egg, but unable to fly, they become tasty morsels for foxes, hawks, and cats, and so they must learn quickly who mother is in order to survive. Within hours at the most, their "mother's" image is blazed into their brains, and from then on they follow that image to safety.

Person-imprinted birds react to people as members of their flock, or as mates, or as rivals. Falco, my beautiful male kestrel, fell in love with John, my former husband (it is too much to ask of a bird to distinguish between the sexes in humans). He also defended John against rivals. One evening when John was helping me with the dishes, Falco flew to the top of the kitchen door, lifted his feathers until he was twice his size, then abruptly took off, folded his wings, and plummeted, talons first. He struck the side of my head like a razor-edged stone, and John finished the dishes alone. Imprinting makes for wonderful but mixed-up pets.

Recalling this, and at the risk of raising an emotionally confused bird, I nevertheless tended and nurtured little Branta,

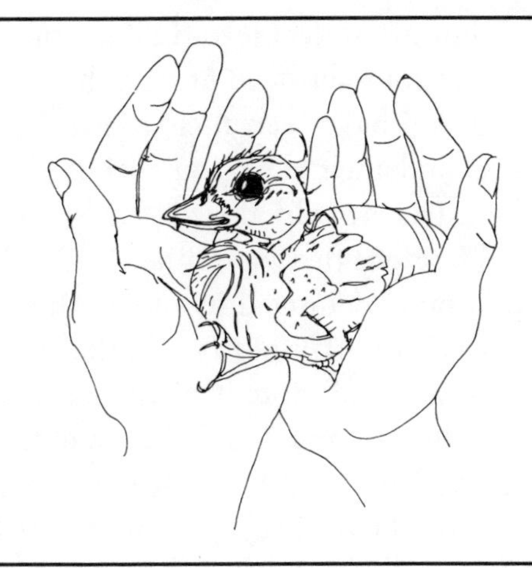

a Canada goose. I had no choice: She was still in her egg when I found her, left behind in the nest when her mother and siblings took off for the water. She and I were inseparable. When I sat down, she sat down; when I went to the kitchen, she went to the kitchen. She ran when I ran, and she stretched up her neck when I lifted my arms to hang the laundry.

Timing is critical in imprinting. If Branta had seen her mother for even half a day, she would have followed geese as well as me and related to both species. As it was, Branta never even glanced at a Canada goose, although they often flew over the house.

*Altricial* birds—like the canaries, finches, doves, and parrots that we raise as pets and the robins, blue jays, and starlings that we nurture after they've fallen from their nests—are hatched naked, helpless, and blind. Safely cared for in nests off the ground or in hollows away from predators, they lead an altogether more passive infancy than precocial birds, without such a strong and urgent need to imprint. Imprinting usually comes some days after hatching, when their eyes have opened. Because they are so feeble and difficult to raise from the egg, breeders do not hand-feed altricial babies until they are well feathered and have seen their parents. The critical period has waned and their devotion to humans is not as strong. The same is true of wild altricial birds, which usually perish if "rescued" before they open their eyes. Hence most altricial birds are not so deranged by our nurturing.

Five- and even seven-day-old robins that I hand raised over the years followed me around my house and yard and constantly kept me in sight much like Branta did. Unlike her, however, they picked up friends of their own species and migrated south with them. They probably mated and led normal lives. Their weak imprinting on humans served them during a fairly helpless period, but it was not so strong as to prevent them from knowing their kin.

If you would like friendly and communicative pet birds of altricial species, you should buy them from breeders who hand-feed them shortly after their eyes open. As for precocial birds, you will either have to make special arrangements with a breeder or incubate a single egg yourself, for they become person-birds only if no others of their species are in evidence. I hatched thirty quail in an incubator and raised them in the house, but not one imprinted on me. They were much too interested in each other and the electric light bulb "mother" that hovered warmly above them.

Fortunately for Jean Bixler's education in bird talk, only Cracker hatched, an accident that has made him a charismatic, feathery person who talks to her in quail talk and to whom she talks back about important human and bird events.

### A BASIC VOCABULARY

The many species of birds, 8,650, makes it impossible to spell out the bird vocabularies in each species' dialect. But there are basic events in a bird's life, common to all, that evoke "talk." By recognizing the events, you can guess the meaning of what is being said and, if you wish, learn how that meaning is conveyed in a particular species' lexicon.

The vocal communications of birds are known as *call notes* and *songs*. All birds have call notes, but only the songbirds have songs. A field guide will clarify for you which are songbirds—they are also called *passerines,* or perching birds, and comprise more than half the birds in the world. But not all passerines sing. Crows, for instance, are passerines but have no song. Their rich sounds are all call notes. You can consider call notes to be the more universal of bird vocabularies and the best place to begin your education.

A bird's own first word is *peep*. Precocial birds like Cracker peep while still in the shell. Until recently it was thought the sound stimulated the mother to attend to her duties, which it probably does, but more important, scientists have found, is that the peeps help the chicks keep their act together. These young—which in short order have to get out of the egg, dry their down, and run along after their mother—must hatch together or get left behind, as was Branta. Peeping coordinates their hatching. One chick upon hearing a nearby chick peeping loud and fast will speed its progress; if it hears only a weak peep, it will slow down.

After hatching, the peep of both the precocial and altricial birds means, when uttered softly, "Here I am" and, when uttered urgently, "Here I am, lost; find me" or "Here I am, hungry; feed me."

The helpless altricial bird adds another statement to the hungry peep. When open, the mouth is a target with a bright red center, rimmed with yellow, that says, "Put it right here." The parent aims and hits. To this add the sight of the wings moving helplessly and jerkily, and the whole creature is almost irresistible. Everyone I know who has picked up a little

*"Put it right here!"*

bird that has fallen from the nest reacts to its pleas by stuffing the peeper's mouth.

The peeped "Here-I-am" call of precocial chicks and ducklings serves to keep the brood together and the mother informed of their whereabouts. As a chick, Cracker ran to meet Jean when she came home from work, greeting her with a flurry of "Here-I-am" peeps. He followed his "mother" Jean like a toddler. He trailed her up and down the steps, around the house, and out into the garden. He confirmed his status as her "child" by posturing: He pulled in his neck. Birds and likewise people hunch up and pull in their necks when it is to their advantage to look helpless. The repairman did so when the by-then-adult Cracker threatened him. Almost invariably, school children pull in their necks and hunch up their shoulders when called before the principal.

Those who turn their ear to bird sounds before the sun is up and, again, as it dips below the horizon can hear the variously spelled chirps, chips, tweets, cheeps, and peeps that, within bird families, add an "All's well," or; "Good-morning or good-night," to the basic message "Here I am."

Parents have special murmurs, soft notes, and warbles that they use within the family to comfort the young. These notes, like Cracker's *chip, chip,* are made deep in the throat and are easy to hear and duplicate. You can tell your bird to be calm, as Jean calms Cracker, by mimicking "All-is-well" comfort notes. What is more, birds like to hear them, just as we like to hear comforting talk from our family. When given at dusk, such a soft sound is "Good-night" as well. Some birds, like the robin, say, "Good-night" with a loud chirp.

At least one nestling call is silent. Altricial nestlings signal with a tail wag that they are about to drop the *fecal bag,* the birds' version of the diaper. Feces are passed inside a tough transparent membrane manufactured by the nestling, which the parent picks up and carries away from the nest to keep the site sanitary and rid it of the white streaks that attract predators. My house sparrow, Tork, told me he was ready to drop his bag by backing up toward me and waggling his tail immediately after he had swallowed his food. I answered by catching the "baggie" in a tissue and disposing of it. This mutually beneficial arrangement went on until he was able to fly. With that, the diaper service ended.

Cracker had no such device, nor did he need it. Able to move around, downy quail chicks scatter their feces across the landscape but, like most birds, at predictable times and places. By observing your pet bird you can figure out its schedule. Birds drop feces upon awakening, before an activity like flight, and before or just after eating. Perry Knowlton, a New York literary agent, would pick up his blue mountain lorikeet every morning just after he awakened and hold him above the toilet. Some birds warn of the coming event by posturing. Falconer Heinz Meng keeps an eye on his peregrine falcon when he takes her visiting. If she tilts slightly and lifts her tail, he whips out a towel and catches the droppings.

## LOUD AND CLEAR

Most communications are vocal, however, especially those that express urgency. A piercing cry clearly says the bird is frightened, and a real screech indicates the bird is hurt or thinks it is about to be hurt. Tork, the perky little house sparrow I raised from a five-day-old, gave his anticipating-pain *tzeeek* just before I sat down on my office chair one morning. He was directly under me, and I replied by stopping my descent in midair.

Birds, like dogs and cats, growl, "Don't come any closer," to fend off enemies. Pet parakeets and cockatiels may also give a hiss if people approach them suddenly or come too close. Tork's growl was *tzerr,* given with his beak slightly open and delivered to my fingers when I reached out to pick him up. Fingers, in his view, were like the aggressive beaks of rival birds.

Parents of some species have a special note, *ick,* to express anxiety over the nestlings, especially when some enemy is threatening them. It is not difficult to evoke an *ick*. Simply walk right up to an active nest. While most birds speak in general terms, some can say specifically which enemy they have spotted. Song-sparrow parents actually have a call that says a person—*pup-puh-puh*—is endangering the young. This worried call is given when you approach the nest and on no other occasion. If you were a kestrel, the song sparrow would say, "*Kruu.*"

Birds that have no specific word for "person" nevertheless have something to say if you approach the nest at all. The first sound you hear is the *alarm.* Of all the call notes, we hear the alarm most frequently. Canaries and parrots use it in the home when a cat or hawk is sighted, as well as when a strange person approaches. Cracker's *squawk-barker,* given when the repairman was discovered in the house, is his species' alarm note.

The alarm grows in intensity as the situation becomes more stressful. When the neighbor's cat comes out to sun, my nesting robins call out their *tclunk* alarm with head feathers raised. Other birds hearing this warning become quiet and look for the enemy. The cat walks near the robin nest tree; the alarm call escalates to a loud *tut* of fear, and now the feathers are flattened, and the neck is stretched. If the cat comes closer, *tut, tut, tut, tut,* issues from what is now a very frightened bird. Other birds vanish. After loud pronouncements the caller, too, takes flight, hides, and pants with beak open. When I hear the call, I answer by stepping out of the door and chasing the cat away.

Some birds are born knowing their natural enemies; most must learn them. Cracker lacked inborn fear of predators. Instead, he has taken on what he believes to be Jean's enemies—strangers on the territory. She was completely unawarc that she was passing this information on to him until he gave the *squawk-barker* alarm call at the sight of the repairman. Cracker is also alarmed by the garbage collector and the electric-meter reader for reasons Jean does not understand.

Realizing that the majority of birds, like Cracker, must be taught their natural enemies by their own parents, I tried to teach Rocket, a robin my children and I raised, that the Airedale Jill was dangerous. When the dog came into the room I imitated the alarm call of the robin, that strong *tut,* which I sort of spit out between my teeth. I even repeated it many times to say, "I'm terrified." One morning Rocket flew toward Jill across the room. *"Tut, tut, tut,"* I sputtered, but to no avail. The bird came on, flying low; Jill leaped, slammed her jaws, and it was over in a second.

On the other hand, some birds inherit the knowledge of their enemies. Zeke, my cockatiel, who was born and raised in captivity and who had never heard his species' alarm cry, screamed a note that raised the hair on my head when a red-shouldered hawk soared past the window.

For most birds, a youngster's first flight is an alarming, though momentous, event that marks its graduation from infancy. I have heard young robins that are afraid to take off give the *tclunk* alarm note. The parents respond with the *tut-tut-tut* scolding. These sounds are loud, but while dangerously noisy, they serve to get the nestlings so keyed up emotionally that they try out their best defense—flight.

Other species pronounce alarm, fear, and, finally panic somewhat differently—*chunk* instead of *tclunk,* or *tik,tik* instead of *tut,tut*—but many songbird species understand one another. Just before the caller reaches the panic-flight state, when the message sounds to our ears like scolding, small birds in the area flock to the scene, adding their own vociferous opinion of the enemy. Tikking, tutting, or shrieking, they descend like a street mob upon the enemy, a harassment known as *mobbing.*

Of the mobbers, robins, jays, catbirds, thrashers, woodpeckers, and red-winged blackbirds are the most aggressive. I should know. When many years ago I would carry the kestrel Falco from the shed to his perch on the lawn, I would hear an alarm from a robin, then from another robin or three. A jay or two would join in, then redwings and the resident downy woodpeckers. All would swoop down upon Falco like a mob of people who have identified a child killer. Sometimes they would knock the little falcon off balance and so distress him that I was forced to take him back to the shed to give him peace from the mob. During the spring breeding season mobbing was intense and frequent, and little Falco was in and out of the shed like the apostles on the Strasbourg mechanical clock.

Finally, there is the *distress cry,* an earsplitting note that instantly sends all birds in the area to shelter. It is the bird equivalent of screaming bloody murder, and it means that the caller is being killed, or thinks it is. Birds hear and react accordingly. Some "freeze" to invisibility, most silently flee. The distress cry will turn flocks of flying birds off their course when they hear it. So predictable is a bird's urge to escape when it hears its species' distress call that ornithologists use it to tell birds to scram.

When starlings became a serious pest in downtown Memphis several years ago, ornithologist Dr. Harry Wilcox rid the streets of them like a Pied Piper of birds. He made a tape recording of the starling distress cry, carried it, a recorder, and an amplifier downtown and placed them under the foul-smelling, noisy roost. He flipped on the switch: *screeeet* pierced the air. An explosion of wings was followed by a terrified silence as the starlings of Memphis departed the area. Wilcox was honored with a citation from the mayor.

"How did you get the tape?" I asked when I visited this longtime friend. He grinned.

"I simply yanked a tail feather out of a live-trapped starling."

Recording distress calls and playing them to birds has

become a standard method of talking bird talk to prevent bird-aircraft collisions. In the 1960s, mile-long flocks of herring gulls began to fly over Newark International and many other coastal airports on their way from night roosts along the ocean, inland to municipal garbage dumps, and back again. The cost of bird strikes was enormous. Windows were broken, fuselages dented, wings and tails damaged. Finally, there was a fatal crash in Boston when airport birds were sucked into a jet's engines. Ornithologists were called in to tackle the problem. The scientists yanked out gull and starling tail feathers, recorded their expletives, and played them back to the birds.

On a morning several years ago at Newark airport, I watched a sound truck mounted with an amplifier drive slowly across the landing strip blasting the herring-gull distress cry to the sky. The gulls, like cattle being herded by cowboys, turned and took another course far from the flight path of the planes. Eventually the distress call so changed the flight pattern of the birds that today they rarely come over Newark airport, and the truck needs only to sweep the sky with sound once or twice a month.

### DINNER BELLS

Searching for food is, in anyone's language, a big event. Among birds, food-search calls are designed to attract fellow birds, for the more birds that join in the feeding flock, the more eyes there are to find food. Some birds talk about food more than others. Titmice, chickadees, finches, and blackbirds constantly announce the search with soft whistles and clinks as they fly through the forest or over the meadows. In winter, flocks of these birds can be located by the noise they make.

When a flock finds a larder, some birds announce the bonanza with yet another note. The starling found-food call is a jerky *creek* that sounds like a rusty gate. If you are a lawn owner who likes to reseed every year, this is a good call to learn and reply to. Last spring my neighbor reseeded his lawn and went indoors for lunch. Within ten minutes a starling sighted the manna, alighted, and called, "*Creek.*" Immediately a second starling arrived, then ten, twenty, almost a hundred. His yard crackled with the call of found food. In desperation he went into the attic, found his grandfather's stuffed owl, and placed it in the yard.

My neighbor had spoken well. Up went the alarm, which escalated to fear, then panic, and the birds took off. But not for long. They learned the owl was no threat and came back a few hours later. Standing in the ruins of his seeding, I told my harassed neighbor about Dr. Wilcox's technique. That afternoon he was in his backyard setting up a live-bird trap to catch a starling. He grinned connivingly as I approached. "Thanks," said he, "this time I'm really going to talk to the birds."

## CHAIRMAN-OF-THE-BOARD TALK

The various "Here-I-am-do-something-about-it" peeps of nestlings are outgrown as birds mature. As baby talk fades, the unique adult calls or songs that repel trespassers and summon mates emerge. Now that Cracker is grown up, he speaks in the language of an adult quail.

His basic vocabulary sounds are not unlike the simple utterances of other species: the contentment *chip-chip* and what Jean describes as a twitter to announce that he has found something good to eat. (Jean responds to his twittered invitation to share a violet bud or caterpillar by running over to see; she draws the line at actually eating flowers and worms, which, fortunately, does not offend the generous bird.) Cracker's *squawk-barker* alarm is also similar to the scolding voice of other species, in that it is sharp, staccato, and repeated again and again.

His territorial call is quite another story. It is the unforgettably distinctive *bobwhite*, known to my generation from the Rinso White soap commercial, and as the once-common name for quails.

Quails don't sing, but their *bobwhite* whistle says exactly the same thing as the songbird's song. As Elliot Howard discovered, the song of a passerine bird is rife with meaning. It announces the bird's species, claims property, advertises for a mate, and is a "name"—an identification sound that tells listeners of the same species who the individual singer is. All this is stated in the glorious "Spring-is-here" song of the meadowlark, the bubbling "Kill-'em-cure-'em-give-'em-a-physic" of the robin, or the cheery *bobwhite* of the quail.

For the first year and a half of Cracker's life he called *bobwhite* softly to keep Jean apprised of his presence and identity. If he wanted Jean to come to him, he would utter the same call a bit more forcefully. He did not belt out the more powerful bobwhite that means "Get off my land," and Jean never expected to hear it, because she lives close to town and rival quails are scarce. One morning she was called from the kitchen by an ear-piercing *bobwhite* call. She hurried into the living room to find Cracker on the windowsill, crest up, neck stretched, staring at something across the millrace that flows past her house. It was a male quail. He was volleying back his strong, loud *bobwhite*, and Cracker was answering, "This is my territory. Do not come any farther or you will have me to contend with. I am I. I am a big, aggressive quail."

All these sounds of Cracker's species, including the *bobwhite* whistle that is analogous to the passerine's song, are inherited. He did not learn them from another quail; he speaks from his genetic code. Cracker's *bobwhite* came from his throat full-blown and perfect when, upon reaching maturity, hormones triggered his inherited announcement.

## CHIRP CHART

*These renditions are based on Margaret Morse Nice's work with song sparrows. Other species have a similar vocabulary, but the sounds are somewhat different.*

| | |
|---|---|
| *see-see, kerr, tit-tit-tit* | Food calls |
| *weech* | Cry of pain, felt or anticipated |
| *eeeek!* (screamed) | Distress call |
| *zhee* (growled) | A threat to another of its own species |
| *tik* | Fear note |
| *tik-tik-tik* (sounds like scolding) | Escalating fear, as with the sight of a hawk or the sudden approach of a cat or person; when repeated over and over, the mobbing call |
| *tchunk!* | Alarm call, also used as a self-assertive statement to a rival |
| *Madge-Madge-Madge, put-on-your-tea-kettle-ettle-ettle* | Male's advertising song, preceded by a chatter when the bird is in flight |
| *kerr* (harsh, unmusical) | Female's self-identification note |

This is not so of songbirds. They must learn at least the details, though not necessarily the musical structure, of their species' songs from the adults among whom they are raised.

Ornithologist Andrew Berger raised a young cardinal, Red, who was imprinted on him and lived in his brick-walled office on the third floor of the Natural Resources Building at the University of Michigan. Red never heard another cardinal sing. He squawked, sounded all the inherited call notes of his species, screamed, and dive-bombed human rivals who visited Berger, but no carol came from his throat. When Berger's

| | |
|---|---|
| *ee-ee-ee* (trilled) | Copulation note and greeting to mate (female only) |
| *tsip* | Recognition note between mates and among kin |

*Some sounds are uttered only at the nest or only by the young before they are old enough to leave the nest.*

| | |
|---|---|
| twitter (uttered softly) | Nest call when hunting for a nest site and the female's nest-building sound |
| chatter | The female's more excited nest-building sound and a greeting to her mate |
| murmurs (miscellaneous) | Parental nesting noises |
| warble | Young's nesting noise |
| *tchip* | Anxiety over the young in the nest |
| *ick* | Stronger anxiety over the young in the nest |
| *puh-puh-puh* | Threat to intimidate a person endangering the young |
| *kree!* | Threat to a hawk endangering the young |
| *eep* (sometimes *ick*) | Young's location note when lost and a worried note before the first flight |

study was over, he banded and released Red only to find it was too late for the bird to learn his species' song. Red was unable to stake out a property and call in a mate. After a few nightmarish days in which he was struck and harassed by a resident male, Berger recaptured him and took him home.

### THE GLORIOUS SONG TALK

Singing is a male achievement; the females of most species do not sing, except—and there are always exceptions in nature—the female cardinal, who sings a few songs, and black-breasted grosbeak females, who purl out songs almost as elaborate as those of their mates.

Still, females, too, learn to recognize their species' song by hearing it as they grow. Both sexes learn their species' unique song or songs in a process akin to imprinting. They are born knowing a few crucial clues for what to listen for and single out the right tunes from among the hundreds of songs caroled on the spring air. As adults, the species' music that females have attended to and memorized is the only music that will induce them to mate, and the only birds singing it are males similarly programmed to memorize those notes. In that way nature assures that they mate with their own species.

Birds, like people, have individual voices, and each mate comes to recognize the voice of the other, just as we recognize

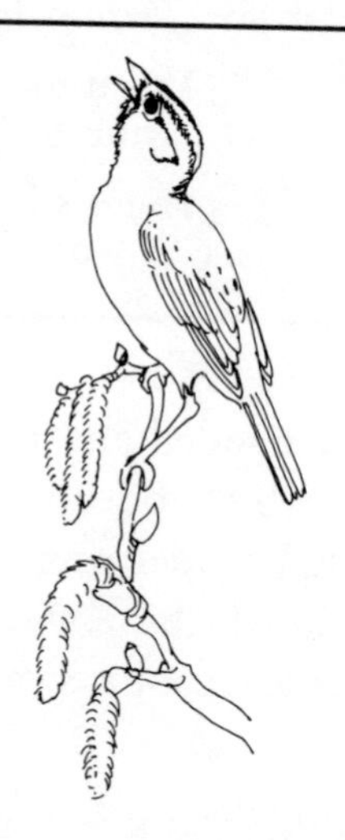

*A male songbird's announcing post*

friends' voices on the telephone. Few of us have the ear to distinguish individual birds by their songs: The incredible song-sparrow specialist, Margaret Nice, could. She knew the voice of almost all the males around her house and each spring rejoiced as, one by one, they returned and caroled out their "names" to say they were safely back from migration.

The song of a bird is not only a name, in the sense that its voice identifies it, but also something that expresses the mood, social status, and vigor of the singer. Females listen to this information when choosing a mate. Needless to say, they wish him to be strong, healthy, alert, and rich. Each male belts out his song from a series of prominent locations within his territory. To determine his wealth in property females note how far he flies to reach his various singing posts. The more separated the posts, the bigger the holding. An inspection of the property is also essential. Females look it over for nesting sites and an abundance of the right plants that will bear seeds and attract insects.

A song sparrow that lived near my camp in a forest in Michigan sang vigorously and exquisitely—but never got a mate. Had I been a female song sparrow I would have succumbed to this glorious singer for his voice alone, but his potential mates were more practical. The species requires the bushy edges of forests, low bushes, and lots of tangles for nesting sites and food. This bird's land was in the woods, a terrible spot for a song sparrow. That little bird just didn't have what it takes.

The exuberance of the male's song is controlled by the amount of testosterone he produces, and that, in turn, is controlled by light in most birds. Listen to your local cardinal or any other male songbird in February. The song is short; his testosterone level is still low. As the weeks pass and days lengthen, the song grows longer, and it is sung more often and with more ebullience. When the complete song is purled out from post and treetop, look in the nest—there will be eggs.

Each species' music is anything from a single note repeated over and over again to a repertoire of compositions that can be so numerous that even our politicians' palabra can't compare. A red-eyed vireo sang 22,197 songs in ten hours, a feat for the *Guinness Book of Records* and a challenge to any filibusterer. As for variety of songs, some birds know more than I do. Male marsh wrens in the western United States have over 200 different songs in their repertoire, and the mockingbirds have many more.

Several birds have songs that are given only in flight, the goldfinch being one of these. The song is sweet and bubbling, a song that says, "I am a goldfinch on the wing." Other species occasionally rise in the air and make rapid "ecstasy" flight songs, markedly different from their regular songs. Ornithologists interpret these air songs as pure ebullience: "I am flying."

### FEATHERY EMPHASES

Some of the least-musical birds, as though to make up for that deficit, are among the most beautiful. The male pheasant courts from a high spot in his territory where hens summoned by his unmusical squawk can get a good view of his shimmering feathers as he bobs and turns to catch the light. They, on the other hand, are drably camouflaged to match the brush, as will be their chicks. The brilliant yellow-and-black goldfinch male perches high on a singing post within his territory where, in case his song left any doubt, females can recognize a male of their species by his colorful feathers. Male mockingbirds perform aerial acrobatics that display their white feathers as they sing. I watched my backyard mockingbird flutter into the air from a tall treetop and, improvising songs as he fell, float on quivering wings down through the flowering dogwoods. His wing and tail markings flashed white in the sunlight until, his fervor spent, he dropped to the ground.

Feathers can tell you about the family habits of your resident birds. Both male and female of those that look alike—chickadees, jays, and titmice, for instance—share the work of feeding the young and usually of nest building and incubating too. Males in species that are sexually dimorphic—that is, species in which the males are brilliantly colored and patterned compared with their mousy mates—do not help with the young, and rarely if ever do they even approach the nest. Instead they sing and flash their colors. They are knights defending the castle rather than nurturers.

The "Come-mate-with-me" meaning of feather displays is so readily understood that few fail to get the point. As an experiment I asked a young man who was watching the pigeons at a city bus stop what a particular feathery bird was doing. Its chest was puffed, and its wings held out from its body and its head, bobbing in and out, emphasized its iridescent neck feathers as it walked among its kind. A strong series of coos rolled from its throat.

"He's strutting his stuff," the young man answered promptly, then added, "Very macho." He grinned. I looked at him again. He, too, was very macho, dressed as he was in red-checked pants and a suede blazer.

Displaying from prominent positions and strutting are only part of the feather vocabulary of courtship. Wings are spread, crests are lifted, and tails are elevated and fanned to reveal beautiful designs to catch the eye of the female. The peacock tail is a beautiful advertisment for love. The birds even know where to stand to enhance their beauty. The pea-

cock on a nearby farm stands in the arched entrance to the carriage house, the sun falling on his iridescent eyespots and throat, making his feathers shimmer against the darkness of the building's interior. He seems to know that the arch enhances the curved beauty of his tail, for he does not stand before the square doorways.

Some birds use feather noises as substitutes for song. The ruffed grouse makes a mechanical sound by drumming. He beats the air with his wings, increasing the tempo until the pulsating sounds come together in a roaring boom. The nighthawk, a skilled acrobat of the night sky, also uses his wings as a voice. As he checks a dive from high above the ground, he lets the air rush through his wing feathers to produce a resonant *whoooom.*

All birds make sweet rustles by lifting and shaking their feathers. The gesture is called *rousing* and is used as a simple "I like you" between mates or among friends. Birds know who their friends are.

When Cracker had a cold in his nose that was making him uncomfortable, Jean took him to the vet, who gave her drops to put in his *nares*—the word for bird nostrils. The maneuver to administer the medicine was unpleasant; Jean

was forced to hold Cracker's head in her fingers and twist it to one side. After the first treatment Cracker sensed she was helping him and thereafter tipped his head to accommodate her.

"Furthermore," Jean said, "he remembered the vet as a kind person too. On the next visit, he roused when the doctor walked into the waiting room."

Most people don't think of birds as cuddly, yet the lifting of feathers, if not accompanied by a growl or other aggressive threat, is often an invitation to physical contact. Mates maintain their bond with one another by affectionate grooming—gently nibbling over the surface of the other's feathers and pulling them one by one through the beak. Sparrow, an unfortunately named Amazon parrot, would present first a ruffled head, then a lifted wing, and would finally flop ecstatically belly-up for a thorough grooming by his owner's fingers. Even more affectionate, from the bird's point of view, is to be the groomer. The same parrot could spend hours pulling his beloved person's hair, strand by strand, through his beak. If a "parasite" is found, the bird will remove it. Unhappily, parrots are under the impression that earrings and buttons are parasites and will remove them in a flash.

### A SHARP EYE

A sense of smell is not one of the gifts of birds or man. Birds are able to discriminate among only about 400 scents as compared with a rabbit's 200,000, and our sense of smell is about 5,000 times less sensitive than a dog's. Little evidence exists to indicate that birds use their sense of smell at all. To find out if my house sparrow could smell food, I tied seeds in a piece of cheesecloth and put it in his feed dish when he was very hungry. He flew to the dish, flicked his tail, gave the food-hunting call, but never pecked the bag. Even I could smell the dusty seeds inside. Tork's eyes, however, were all seeing. One small hole poked in the cloth with a needle and he whacked open the wrapping and stuffed himself.

Bird vision is so acute as to seem otherworldly. I watched a golden eagle in Idaho that was soaring a mile high, turn, plummet, and come down upon a ground squirrel with uncanny precision. You need not watch an eagle to be impressed by bird vision; just watch one of your local songbirds fly through the treetops at ten to twenty miles an hour without hitting a limb, leaf, or twig.

The sharpness is accounted for by areas called *foveae,* spots where photoreceptor cells are packed tightly together to form a point of sharpest vision. We have one fovea at the center of each eye's retina. Most birds have a central fovea plus a second one to the side in each eye. The central ones are used together as they are in us when we focus on an object. But unlike our visual field, which is precise only at the center, a bird's-eye view includes an additional area of sharpness to each side, corresponding to the two additional foveae. As the bird flies, objects in its view are seen clearly from three angles simultaneously.

Such precise vision can be terrifying. Andrew Berger's cardinal, Red, defending his territory in the university office, often flew at the eyes of his human rivals. Most fended him off. One unaware person, however, was hit. The strike was so fast that he had no time to blink. And it was so accurate that only a creature with exceptional eyesight could have done what he did. According to the optometrist who treated the man, the bird struck the pupil dead center and so controlled his strike that the beak only penetrated the outer membrane, a microthin skin. Incredibly, no harm was done.

Imagining three-field eyesight is beyond me, but research

is deciphering the bird's-eye view. Experiments show that birds see and react not to the whole, but to the part. A robin reacts to a red breast. It says to him: male robin, intruder; fight. Evidence that it is the red breast and nothing else was demonstrated when an experimenter painted a ball of cotton red and staked it out on a robin's territory. The resident male dive-bombed and tuk-tukked at the piece of cotton. However, when a live male robin was dusted with white powder and tethered out on the same property, the resident male did not even sing his territorial song. He completely ignored him.

To Tork, a rival house sparrow was a beak. Anything that even vaguely resembled a pointed beak was attacked—pencils, fingers, nails, sticks. I discovered this when I sat down, pen in hand, to do my monthly accounts. Out from his roost in the bookcase he would come, fly directly at the end of the pen, and crack it numerous times with his beak. When I put it down, all aggression ceased. For some reason, it no longer

resembled a beak. When I picked the pen up, he attacked it again, and I was hard-pressed to get budgets balanced and checks written. In the spring of the year he became so agitated by pens and pencils that I was forced to work in my bedroom behind closed doors.

You can take advantage of birds' dramatically programmed perception to attract hummingbirds. Pour sweetened water into any small container—test tube or medicine bottle—decorate it with a bright red ribbon, and wire it to any plant in the garden. To the hummingbird's strange vision, a bow is the same as a flower, and it will answer your advertisement by whirring in to sip as day fades to evening.

Fascinated with Tork's response to pointed shapes, I repeated a famous experiment on sight cues. I cut out of cardboard a silhouette of a soaring hawk reduced to its simplest abstraction: a cross, in which one bar is the wings and the other the short head and long tail. I suspended it on a string and swung it over Tork, small end forward. Tork gave the alarm-call note; he had seen the enemy. A few days later, as other experimenters had done, I moved the cross long end first, which made the cutout a long-necked image—a flying swan or goose, a nonpredator. Tork gave no response; he went on feeding.

Bird-watchers, after a great deal of experience, become very good at recognizing species by the flash of a patch beneath the wing or by the bar on a tail. With apparently no effort at all, birds recognize the differently patterned males, females, and juveniles of their own species. Small details of the patterns have recently been discovered to serve more specific purposes than species, sex, and age recognition. The mallard, the most abundant duck in the world, has a violet-blue patch on the wing that says, "Here's mother." Ducklings that stay close to that signal usually survive. Those that drop

*The cross as a dangerous hawk*

*The cross as a harmless goose*

back and lose sight of the "mother spot" are prey to hawks and snapping turtles.

When the young grow up, parents may remain with them in family groups and may also gather in flocks of many families. Markings flashed to mates in courtship are now used to keep birds of a feather flocked together. "Follow me," the white mark on the tail of the junco says when one takes off from the ground in a hurry. Members of the flock see the white flash, get to their wings, and follow, usually away from danger.

## METAMESSAGES

Students of bird behavior acknowledge three patterns of bird communication. The first is the *discrete message.* Discrete messages are the most dramatic in the bird vocabulary. They include territorial songs, call notes, and "one-word" postures, such as the herring gull's neck-stretch stance, which declares his status. *Compound messages* are those that intensify the feeling expressed in a discrete message by combining it with a visual display or gesture, such as when a pigeon coos and also struts his stuff. The herring-gull male postures his discrete status message, standing tall with his neck out, when another male approaches. If the rival does not depart, he compounds his statement by leaning over and pulling grass to emphasize his aggressiveness and his ownership of the property. The added emphasis is comparable to a man asking a stranger to depart from his land and, if he doesn't, shaking his fist in his face.

The final pattern, laboriously labeled *metacommunication,* tells the hearer how to interpret the message by putting it in a context. A crow can use a threatening-call note but tumble in play to say its first statement was just in fun, like Charles Darwin's dog who wagged his tail while pulling on the scientist's sleeve and growling ferociously.

We use the term "Birdbrain" to express our doubt that much intelligence can reside in so small a container. In fact, birds do not have room in their skulls for the large, heavy cerebral cortex that is the seat of human learning and conscious thought.

Yet the complexity and even the originality of bird talk is startling. Crows, for instance, crack jokes. My pet crow, Crowbar, would bring me an acorn while rolling out his "Share-the-food" call—and the acorn would be empty! In 1983 it was discovered that birds' capacity for vocal learning—and presumably the crow's wit—lies in the hyperstriatum, which is analogous to a part of the human brain notable in us only for coding automatic behavior. Those philosophers and scientists who had thought learning and intelligent behavior was impossible without a large cerebral cortex had to think again.

Certainly those of us who own birds believe that they can think. Researchers have substantiated that intuition. Birds, they have demonstrated, not only learn their songs but can solve all the usual forms of mazes and problem-solving devices used to test the intellectual abilities of animals in laboratories. Star, a great tit native to Europe and similar to our chickadees, was raised by British naturalist Len Howard, who taught her to count to 8 in response to his verbal request. Although Star lived outdoors, she daily came into the house to visit Howard. "Tap 3," he would say to the attractive blue-and-gray bird. She would tap 3 times, then look to him for another number. "Tap 8," he would say to her. "She gave 8 in 2, 4, 2, a new division of her own invention. She repeated it twice as if enjoying it, and was praised with a nut," Howard said, "6"; Star tapped out 3, 3, then 2, 2, 2. "She had a precision in her manner of tapping," Howard wrote, "and never hesitated after being given the words only."

Birds not only have the ability to count, they also have excellent memories. Clark's nutcracker, a crow and jay relative that inhabits the Southwest, in the fall stores up to thirty-three thousand piñon-pine nuts, four to five seeds per hole, in thousands of hiding places over many acres of land. For the next six months, it is able to recall nearly every cache, the locations of which it has memorized in the form of a mental map. If landmarks, such as boulders, are moved, it can no longer find its stores.

A pair of robins on a Michigan farm were upset by the experience of being leg banded for identification. Thereafter they dive-bombed and screamed "Enemy" whenever the bander, my former husband, walked under the nest tree. What is more, they mobbed him every spring for three years when they returned from migration. I could walk across the territory without being attacked, as could every other person. John, they remembered.

**ADDRESSING THE CROWD**

Birds have been domesticated for at least five thousand years; the first record is chicken remains found in a midden that age in China. It is not surprising that the first to join the human family was, like Cracker, a precocial species.

Knowing what we do about imprinting, we can imagine that the chicken ancestor, the jungle fowl of Southeast Asia, was induced to stay with us in much the same way that Jean socialized Cracker. An egg brought from the jungle for food might have hatched in a human hand and the chick imprinted on its person-parent, giving villagers the bright idea that perhaps groups of chicks raised among people would stick around.

Over the eons the wild bird was altered by breeding, first as an ornamental pet, then for the sport of cockfighting and the practical production of eggs and meat. The original jungle fowl is as different from the Plymouth Rock meat chicken as an ear of corn is from its progenitor, the tiny seed head on a blade of tripsacum grass, which still grows in Florida.

Over the millennia, other precocial birds that imprint easily—geese, ducks, swans, guinea hens, and peacocks—came to live in farmyards and parks.

Keeping flocks of birds, as opposed to keeping a single pet, requires some understanding of their social lives. If you buy a group of chicks, for instance, you have to face the fact that as they mature, you will have to cull from the flock all but one rooster. Chickens are polygamous, with one dominant rooster defending a harem of hens and fighting furiously with any other rooster who tries to horn in.

Geese, and many species of ducks, on the other hand, are monogamous and mate for life. A bonded pair suffers deeply if separated. A suburbanite went to buy his child a nice white goose as an outdoor pet from a local pet store while the owner was vacationing. Not knowing the deep bond between mated geese, the fill-in salesperson broke up a pair to sell the man a gander. The gander could not settle into his new home. He called piteously, his feathers became unkempt, and his head hung low. Finally he refused to eat. Fortunately for the gander, the store owner returned to find the goose behaving likewise. He put her in his car, drove her to the gander's new home, and released her. The reunion was a striking sight as they rushed together with uplifted wings and stretched-out necks, uttering the vocal goose greeting, "*Uh-un, uh-un.*"

Some domestic male and female geese look alike, but mated pairs can be discerned by observing their behavior. They stay close; if one moves off, it checks on its mate's

whereabouts frequently with eye contact, albeit that peripheral view, which to us does not seem directed at the mate. If you feed the pair, one will stand guard, neck up, while the other eats, then vice versa.

If you want a flock of geese, it is wise to either buy them as a group of goslings or let a mated pair produce their own flock. Geese do not take to strangers of their own kind or, for that matter, strange dogs, cars, or people. For that reason, they have for centuries been kept as watchdogs. With loud clarion notes geese announce trespassers, often before dogs are aware of a stranger's approach, and attack without hesitation. Although they are not killers, they hit with their wings with such force that they can break an arm, and they can bring blood when they bite.

In purchasing homing pigeons, the selection of a pair is even more important, for their love becomes a language through which owners can teach them to home. The female is kept in the wire roost cage where she remains visible. The male is released. He explores the environs on wider and wider excursions, but he always keeps an eye on his beloved in her cage and always returns to her. After a week or so, the female is released to learn the lay of the land from her mate—and the route home.

### SPEAKING OF HOME

Shortly after hatching, birds, both altricial and precocial, are imprinted not only on parents but on place.

Place imprinting brings a yearling robin back in spring to the same type of habitat in which it was raised—sparsely wooded barrens that translate into yards and forest borders in modern times. The returning young bird moves off about ten miles from its birthplace, a shift that avoids incest, but it is so strongly imprinted on its natal surroundings that it often chooses to nest in the same species of tree as the one in which it was raised. Apple orchard imprinted robins, banded so that they could be recognized as individuals, tended to come back to apple orchards, backyard-imprinted robins to backyards.

By making use of place imprinting, biologist Fran Uhler of the U.S. Fish and Wildlife Service was able to induce black ducks of the Patuxent Wildlife Refuge, Maryland, to change their nesting habits and thereby save their lives. Raccoons were destroying so many black-duck nests and eggs that the species was in danger of being wiped out. One spring Uhler put mother and eggs into raccoon-proof nesting boxes. Faithful to their clutch, the female ducks stayed with the eggs and incubated. When the ducklings hatched, they saw the inside of the boxes. The following spring those young reversed a million years of family tradition and nested in boxes.

Cracker learned his territory—his borders, where the food is, and where he can hunker down—by following Jean as a chick. As far as he is concerned, his home place is inside the redbrick house on the corner of a quiet street in Boiling Springs, Pennsylvania. Could he select a spot for his own family, he would, no doubt, pick the inside of a house on a quiet street in a small town.

By understanding that birds imprint on the type of landscape, species of nesting trees, and kinds of food plants to which they were accustomed as youngsters, one can also bring birds into a place they would not otherwise have chosen. The trick is to make the place—it need be no more than a backyard—"familiar" by planting it to suit their preconceptions.

No one was a greater authority on summoning birds from the wild than Herbert Johnson, a horticulturist specializing in roses and petunias who was assigned by Robert Moses, New York City Parks Commissioner thirty years ago, to transform a dump at Jamaica Bay into a wildlife refuge.

The site was not promising: a desolate landscape of bedsprings, refrigerators, tires, and excavated rock and mud, visited by no more than fifteen species of birds.

Johnson effected the land change on the eighteen-square-mile tract in the boroughs of Brooklyn and Queens by simply watching what the birds said about their preferences.

"I knew nothing about birds," he said. "I was a flower gardener, my father worked for the queen in Kensington Gardens and never taught me a bird, just flowers. But I have eyes. One day I saw lots of birds in a field eating weed seeds. I called Mr. Moses and told him to have the city trucks unload the topsoil they were carting away on my barren dump. Topsoil, I knew from experience, is full of weed seeds. The weeds popped up after the first rain, flowered, seeded, and the birds came.

"I also knew birds like water by watching them in the city parks. I dug two freshwater ponds, and after a while ducks came to the water."

Johnson then walked the nearby Jones and Coney Island beaches and collected trees and shrubs that grew in that salty environment: Japanese pine, Russian olive, and bayberry. He transplanted them to the similar environment of Jamaica Bay. On a trip to Connecticut marshlands when the park was three years old, Johnson saw a flock of ducks drop down to feed on eelgrass. He went home and planted the water as well as the land. Into his eelgrass came snails, crabs, shrimp, and killifish, followed by plovers, sandpipers, willet, dowitchers, phalaropes, terns, and ibises—birds that had not been seen in the New York City area for many, many years.

After five years, the early plantings had matured. That fall of 1958 trains from Jamaica Bay to the city were brought to a halt when five to ten thousand swallows descended low over the rails to eat the seeds of the bayberry bushes, which many other new species were already enjoying.

"When the Russian olives matured," Johnson said, "they attracted thirty-five more species."

The Japanese pines grew and attracted owls. "The first was a little saw-whet from the North," Johnson said. "The newspapermen became interested in him, photographed him because he was cute, and he appeared on the front page of several papers. Hundreds of people came from the city to see him. They are nice little birds. They just sit there and stare at you quietly."

The saw-whet was followed by the arrival of great horned, snowy, barn, and short-eared owls.

"That little owl got me really interested in planting special plants to get special birds," Johnson said. "I planted thistles, and down came goldfinches; raspberries, and in came mockingbirds; sea oats, and the rare seaside sparrow appeared. Wild cherries brought the myrtle warbler. The shadbush called in the bluebirds and the thrushes, and the linden trees brought the rose-breasted grosbeaks."

In May of 1959, Herb Johnson and his plantings made ornithological history. A glossy ibis, a nearly extinct bird of the southern swamps, appeared in the sky above Jamaica Bay, dropped its landing gear, and settled down to stay. A few months later a white pelican, a bird of the Far West and the Gulf of Mexico, took up winter residence at the refuge—the first time one had come to stay in the area since 1895.

Word spread. Bird-watchers and ornithologists from all over the world visited the refuge. The roster of annual visitors grew from 35 to 100,000, the birds from 15 species to 315. The area had become a rendezvous for people and birds.

Before his death in 1981, Johnson recalled his favorite planting. "I had just planted a hemlock tree someone had donated," he said, "and had stepped back to admire it, when down came a robin. She began building a nest right in front of me, then and there. I believe she saw us planting that tree and waited until we were done." He shook his head. "It's uncanny how it works."

### TALKING BACK—BUT HOW?

There is a fantasy shared by many people that one day, in wood or meadow, they will somehow say or do the right thing, and wild birds will understand they are their friends and will come to them.

Jean Bixler is able to bring Cracker running to her side because his vocabulary is small and because her vocal cords and whistle can closely imitate his sounds. But birds don't make sounds with vocal cords as we do, nor do they whistle.

Birds produce sound through an organ called the *syrinx*, a group of vibrating membranes surrounded by an air sac and located where the windpipe forks into the two bronchial tubes that carry air to the lungs. The windpipe itself may be very long and twisted into coils like a trumpet. The honk of the goose, the trumpet of the trumpeter swan, and the whoop of the whooping crane owe their loudness to coiled windpipes. The males of some species amplify their calls further by an additional resonating chamber that pouches out from the windpipe and gives their sound a booming quality. Apparently the two bronchial tubes can toot separately through the syrinx. An exasperated mother and parrot owner, besieged by her children's demands for whistles but knowing only too well that the noises would be instantly and incessantly repeated by the bird, thought she had found the solution: a whistle that produced two harmonizing notes simultaneously. The children were delighted. So was the bird. He listened once, and tooted—two notes at once.

Depending on the construction of their particular instrument, birds also produce throaty growls, metallic clicks, hiccup sounds, clear tones, rough rasps, wet burbles, and glorious melodies—almost all of which are difficult for people to imitate.

A few are not. The chickadee's *hi-sweetie,* the cardinal's *cheer, cheer,* and the tufted titmouse's *peter. peter. peter,* as well as the quail's *bobwhite,* can be whistled by a moderately good whistler. Numerous people have learned to imitate these songs by listening to the Cornell University recordings of birdsongs, available through the ornithology department. Done accurately, these announcement calls will bring a bird to within a few yards of you as it checks out the new "bird" on its territory. Finding only a person, the individual will linger, eyeing you as if to make sure there is not a bird in your pocket.

Sounds not possible to imitate by whistling can be twisted out of an Audubon birdcall, a cylinder that is pressed and turned inside its case to make chips and clicks. Call notes, those simplest sounds, are the easiest to imitate. Chipped feeding calls can attract certain birds during flocking seasons,

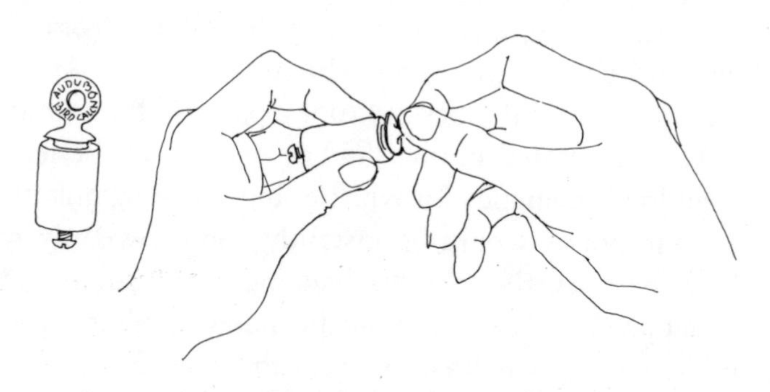


*The Audubon birdcall*

and at all times alarms will do the opposite. Some experts can produce musical sounds on this gadget. A friend of mine twisted and turned until he reproduced a very good red-winged blackbird territorial song—and brought down upon his head a screaming male bird who was protecting his harem. I have heard that some people have summoned with a twist of the cylinder bluebirds and even the trilling meadowlark.

My local pet-shop owner swears by a homemade device—a cork twisted in an empty wine bottle. He not only calls birds out of the field near his store with a bottle and cork, but uses it to teach untutored male canaries to sing. Kissing the back of the hand works very well for me. I can duplicate quite faithfully the peep of lost babies and the alarm calls of some species. Done swiftly, kiss noises sometimes sound like mobbing, for last winter I gathered both chickadees and nuthatches with squeaky kisses to the back of my hand.

The most effective device, however, is a tape recorder. A wood thrush comes to the opening in the forest on the hill behind my house when I play back a recording of his voice. He looks for the rival and threatens him by purling out his wondrous song in full view.

"Owl calls," on the bird call cassette that accompanies the Peterson Field Guide Series, which can be ordered through your local bookstore, changed the hullabaloo of a junior-high camp-out into silent reflection. A teacher suggested to a tentful of boisterous boys that they accompany him on an owl walk. About one hundred feet down a dark trail he took out his recorder, slipped in the owl cassette, and flipped the switch. He had taken off, and recorded over and over again, just an owl call. The soft wavering call of the screech owl floated out through the forest. The boys stopped talking and listened. From far away came an answering call. The cassette played on, the distant owl came closer.

"Then he sat right above us," a boy said. "A little owl with huge eyes and a big head. He thought we were owls. Then he saw we were not and flew off into the darkness. He was awesome. We tiptoed to the tent and kept quiet, hoping he would come back."

## CALLING ALL CROWS

With or without gadgets, the bird you can talk to most easily is the crow. Farmers and hunters ages ago broke the language code of these birds, which ravage crops and steal eggs, in order to call them into view and shoot them. They did it with their own voices.

The highly intelligent crows have the most advanced communication system of all the birds. Through it they talk to each other, announcing their mood, the approach of a predator, the location of food, or a cause for assembly. Crows can even tell each other when food has been poisoned. In the 1930s the U.S. Department of Agriculture recommended that farmers rid themselves of these pests by treating corn kernels with death-dealing poison. The ruse worked well at first. Crows ate and dropped dead. But a dead crow is all crows need to see in order to start spreading the word. A warning went out: "Avoid the poisoned grain." The information was specific. They avoided the treated corn but ate untainted corn in the same field.

Crows in my neighborhood perplexed me with their intelligent communication. I could not understand why I never saw them opening the garbage bags or stealing the corn I threw out for other birds. But I knew they did these things. In winter their footprints were clearly marked on the snow, and their beaks left identifiable holes in the bags dumped out of the metal cans by raccoons.

Many mornings I got up early and sneaked to the window hoping to catch them in the act of scavenging, only to look down on an empty and innocent-looking yard. Then it all became clear.

Lying in bed one dawn, I became very aware of the crow that sat on the branch of my linden tree every morning. He was, I realized, staring at something in my room. Slipping quietly out of the far side of the bed so that he could not see me, I inched toward the spot his eyes were upon—Jill, the Airedale. Every morning as I stirred from sleep, Jill arose in anticipation of being let out—and unbeknownst to me until that morning, the watching crow sent out the "Enemy!-Here comes-the-dog" cry the moment Jill arose. The crows at the garbage and bird feeder lifted their wings and silently left the yard before I could have seen them had I followed my usual routine and a good minute before Jill could have chased them. They watched and knew my habits—an eerie thing to contemplate.

*"Here comes the dog!"*

Although farmers and hunters passed on to each other the language of the crows for centuries, it was not until 1952 that the definitive translation of crow talk was set down by scientist Eugene E. Good. Two other scientists, D. R. Chamberlain and George W. Cornwell of Virginia Polytechnic Institute, further discovered postures that accompany some calls.

For those who want to speak to these smart and entertaining birds, the famous Herter's Inc. "Complete Crow Call Manual and Hunting Guide" focuses on those calls that, used in combination, will bring them in close. The manual was written to be used with the wooden crow call, a three-inch cylinder with two wooden reeds that is blown like a clarinet and can still be bought at sporting goods and hunting stores. Although the directions were written for hunters, photographers and bird-watchers are now able to call in crows from these directions.

In uppercase letters the manual warns against the indiscriminate use of the crow alarm call and says never to use the distress signal.

Good describes the alarm that warns of a predator as *ca ca ca,* a short staccato cry with no change of pitch. The cry is accompanied by wing flicking, head bobbing, and beak snapping. The crows react to this warning by becoming quiet and often flying silently toward the source of trouble, but under cover where you cannot see them.

The distress call, which announces disaster, as when a crow has been shot or poisoned, is *CA CA CA CA CA,* five sharp, loud cries trumpeted hysterically.

"This call will drive crows away," Herter's manual reads, "and will alert every crow in the entire area in no time." Of course, if you want to protect your corn or birdseed, that may be just what you want to say to the crows. The manual adds that to learn this call all you have to do is "frighten a crow (shoot toward it or merely let it see your gun) and listen to the high-pitched fast tempo of the call."

Another cry described by Good is *CACACACACACA* repeated over and over again in loud desperation. This is their mobbing call, which assembles nearby crows to dive-bomb the enemy and which Good also translates as "come harass the owl." The cry, emphasized with bobs of the head and flicks of wings and tail, is the most distinctive of all the crow calls. It halts other species of birds and sends them to cover, alerts squirrels to the enemy, and brings binoculars to the eyes of knowledgeable bird-watchers. The manual advises, "Keep this call excited. Like most crow calls, this one is given in a series of three parts. Blow the call until you have the crows definitely on the way." Now, the manual advises, it is time to use the alarm *ca ca ca,* the same short staccato call the crow outside my window gave when it saw my dog get up, in hopes that the crows will be encouraged to silently steal closer to the source of trouble.

More must be done; these birds are clever and might at any minute detect the sham: "As the crows come toward you, you must mix the FIGHTING CALL and the RECOGNITION CALL." Recognition, translated as a slow *ca-ca-ca* by Good, is supposed to reassure the now somewhat worried crows that they are not alone. The fighting call *ca-aw, ca-aw, ca-aw,* a long guttural rattle, is a threat akin to the growl of other passerine birds. It announces a crow's intention to take on whoever is threatening its place, food, nest, or young. Its use at this point, the Herter manual explains, is to emotionally gird the crows for battle. "Keep it up until you shoot," concludes this colorful writer.

Good, interested more in what the crows are saying than in calling them to the gun, has additional words in his crow dictionary.

The most common call of the crows, he found, is *ca-a-a-a,* descending in pitch in groups of two to six. It is a social formality; it says who the bird is, how it feels, how the weather is, or some other thing no crow can do anything about. None turns a head at these remarks.

*Ahh, ahh, ahh,* is a wail usually given by a captured crow: "I am in the hands of the enemy." The wail summons crows to the rescue; they hide in tree canopies or behind limbs and trunks to observe and to plot a rescue.

*CACACA* sounds something like the assembly cry "Come harass the owl" but is given by parents in spring when an enemy is approaching a nest. It is therefore equivalent to the song sparrows' *ick,* meaning danger to the young. Upon hearing the cry, unmated birds will arrive to help battle the intruder.

Although several crows will nest in the same pine stand with one another, it is during the nonbreeding season that crows by the thousands assemble for the night at communal roosts. The large group breaks up into smaller feeding groups as they spread out across the countryside in the morning.

*Ca cac caaaa* means a crow is flying to the roost, *ca cac car* that it is leaving the roost to feed. Again other crows simply listen to this. They do not react. It's like your saying, "I think I'll brush my teeth." No one usually runs out of the house and buys himself a toothbrush to join the activity. You might, however, say, "Okay," which is approximately what the crows say when they answer with a similar cry.

Three or four pairs of short simple *CA*s are given when mates are apart. These are adult location signals and are called out only when the birds are perched. They are answered with identical notes.

The most mysterious and provocative of all the crow cries is *nevah, nevah*—the call of the dying crow. It is a raucous sound, uttered slowly. Other crows react unpredictably; some stay, some go, some wait and hold a funeral.

The crow funeral has been seen by very few people. Those who happen upon it know instantly what it is by the expression of grief. A friend, who is a good observer, described with no little reverence the crow funeral she witnessed along a country road.

"When I approached, the birds were sitting on the limbs of a tree above a dead crow. Their wings and heads were drooped," she said. "The scene was somber. The crows were sitting together in twos or family groups. When I arrived, they were utterly silent; then, as if realizing the finality of death, they moaned sadly as if their hearts were broken, like people at a funeral."

Crows also coo and rattle. The coos and rattles are given year-round according to some observers, but in my experience they are associated with mating and pair selection. While walking in the spring woods, I heard a strange belling, then a rattling as if someone were striking hollow sticks together. Sneaking toward the sound, I saw two crows tumbling from limb to limb, fanning their feathers, and speaking of love in these strange, dark sounds. The two floated to the ground, where one pressed its belly to the earth near the other and, spreading its wings, slowly toppled over as if in a trance. I did not see if they copulated—bird copulations are over in an instant—but they presently flew off to a nearby branch and roused, "I love you," as birds are wont to do after mating.

## STRAIGHT TALK

Scientists agree that birds talk to one another; hardly any agree that they direct their comments to humans. I claim—though scientists would shake their heads at this—that my wild birds talk directly to me. Every September my chickadees sit in the trees outside my window and, looking straight at me, scold to tell me it is time to fill the feeders with sunflower seeds. I answer them by doing just that. Satisfied, they stop berating me.

On one occasion, a pair asked me for help. A cat was stealing toward their nest tree. With a whir of wings, they came to my porch door and began clicking to me through the screen the "Danger-to-the-young" call. There was no question in my mind but that they were telling me to get rid of the cat. I hurried out, picked her up, carried her to her mistress, and suggested she keep her cat inside during the breeding season—quite a flurry of action after a few emotional clicks from two little chickadees. Would that English were as sparse and effective. I have spilled out cascades of words to my children about picking up their clothes, and nothing happened. One good click is worth a thousand words, I say.

Purists tell me that my chickadees were clicking to state their anxiety, not to tell me anything. (But why did they direct their comments at my screen door? And why did they eye me?) These critics almost convinced me until Dr. Steve Zack, an expert on a family of African birds called *rollers,* paid me a visit. I told him my experience with the chickadees and asked him if he thought birds ever deliberately communicated with people.

"Have you ever seen a honey guide communicating?" he asked. I had not. But I had heard a yarn about the greater honey guide, a grayish brown bird about eight inches long with a bright pink bill that, the yarn went, led human beings to the hives of wild bees.

"Isn't that just a myth?" I asked.

"No," he answered. "It's true. We have learned it's beneficial to both people and birds. The people eat the honey and leave the wax and larvae for the honey guides. For some reason, the birds need beeswax in their diet." He paused.

"I had the honor of being picked out of all the people in our camp in Nakuru National Park to go with the birds to a bee tree.

"A pair of honey guides had been flitting around the camp for several days to the pleasure of everyone. One afternoon as I was walking to my tent, the birds alighted on a branch above me and began chattering excitedly. I stopped. They looked right at me, calling, '*Ke ke ke ke*,' over and over again. I had never heard a honey guide call a person, but it was perfectly clear to me what they wanted. I walked toward them. They flew to another tree and called again.

"I did not follow. They came back and called. I went with them. In this manner we progressed through the forest—they calling, flying ahead, coming back to get me, and calling again. I was thrilled to be the person selected for this assignment and followed in awe.

"Occasionally I got off the trail on purpose to see what they would do. They found me, called frantically, and led me back on course.

"After a long walk they alighted in a tree and fluttered their wings excitedly. They said very clearly, 'Here it is.' I circled the tree, and there was a beehive about twelve feet up."

## MISUNDERSTANDINGS

Every bird owner is convinced that birds purposefully state their wishes to people. The problem is accurate translation. When my ex-editor and friend Andy Jones moved to Santa Fe to become a writer, I inherited his cockatiel, Zeke.

As I learned cockatiel talk, jotting it down and whistling experimentally, I learned how little I really knew about bird speak. Zeke had been very attached to Andy and had great difficulty adjusting to me. Not only did he scream, "Enemy!" when I came into the room, but he pulled and twisted his feathers in anxiety.

I set aside time each day to win his affection, usually talking to him by imitating the whispering nest sounds of the loving parent. He calmed down, screamed less, ate heartily, and eventually sat on my shoulder. A year later he sang, "*Krickity, krickity, krickity,*" the first part of a three-part love song of the cockatiel. I could imitate this rather well and would duet with Zeke in full whistle. Duetting, in which two birds take turns calling phrases to one another, binds pairs together. In some cases, one bird utters an incomplete phrase, and the other completes it so perfectly that the human listener cannot tell that the phrase was made by two voices. I hoped my duetting would lead to the bird equivalent of an engagement. It seemed to be working. Several days later Zeke sang the second part of his love song, then with a whistling crescendo sang out the third. Our relationship had flowered.

And yet, he still yelled, "Enemy!" when I approached him. One day I realized he had discovered his reflection in the downstairs powder room mirror and was being held captive by his own image. He stared, lifted his feathers, and sang the first, second, and third parts of his love song to his reflection. A few weeks later he discovered his image in the glass on a painting in the living room, then in a silver platter. One afternoon I came home to find Zeke building a nest in the dark space between the phonograph and the recorder. He was filling it with bits of paper torn from discarded manuscript in my wastebasket.

His interest in me grew. Free to fly about the house, he greeted me with swoops, dives, and *preet* greetings. He circled

me when I walked to my study and, at last, as I entered the powder room one day, flattered me by alighting on my head and fanning his wings and tail.

One morning he rode my shoulder from my word processor into the powder room, where I happened to glance at him in the mirror. His tail and wing feathers were fanned beautifully, I thought, and now his crest, too, was erect. I had never seen the feather displays of a courting cockatiel, but surely this was it. Our problems were over. Zeke would now treat me as lovingly as any cockatiel treats his mate.

Andy called that day and asked about his friend.

"We've finally hit it off," I replied happily. "He's courting me and building a nest for me. He sits on my shoulder and dances on my head. He swoops down to show off when I walk into the room. He even whistles the wolf call you taught him."

Andy was pleased; I hung up the phone and went into my office to change the cassette on the tape player. Zeke was admiring his image in the powder room mirror. I was near his nest. From afar he screamed the enemy cry, flew at me, and hit my head with a nasty wallop. Then he climbed above me and dive-bombed again and again.

I faced the truth.

Zeke was not in love with me. I was his enemy, a threat to his home between the recorder and the phonograph—the home he intended for his mate in the mirror, the picture glass, and the silver platter.

Zeke was not courting me at all but battling me in defense of his imaginary family. I tested my new theory by walking up to his nest. He pounced on my head and bit. When he sat on my shoulder, he did not rub his head against me as he did to the image in the mirror but bristled his head feathers and crest in anger. I ran another experiment. I whistled his morning greeting, "*Preet . . . pre-e-et.*" He flew to the mirror to answer. Zeke was talking to his image bird, not me.

The soft *pe-or* called at dusk, which means "Good-night" and which I thought we had learned to say to each other out of love, was directed not to me but to the bird in the silver platter. Our morning conversation, which took place as I sat down to the word processor, was of war, not love.

**PLAIN ENGLISH**

I wished, as the pretty gray cockatiel went back to his mirror love, that I could explain myself to him in my own language, since I had translated his so badly. Cracker knows a little English. He comes when Jean calls him by his human name. "Come here, Cracker" brings him from what he is doing upstairs, hopping-flying down the steps and running across the floor to her feet. He also knows "Let's go to the garden." This will bring him from the windowsill in the living room where he now watches for that threatening bobwhite neighbor, or it will rouse him from a snooze in his nest-box. Furthermore, he talks back to Jean in human lingo. He shakes his head "No" when he is offered food he does not like. The first time Jean gave him a sticky garden leaf, he moved his head from side to side to avoid it. She withdrew it. The next time she offered him a sticky leaf, he repeated the successful gesture more adamantly—from side to side went his head "No, no, no." Now it is part of his vocabulary.

The temptation to conduct a conversation in one's native tongue is most compelling with those species that mimic us, for they are very convincing. My pet crow, Crowbar, could say "Hello" and "Hi ya, babe" so clearly that on occasion I hurried to the door to see who was paying a visit. A captive crow at the Massachusetts Audubon Nature Center sent people into gales of laughter by shouting from its cage, "Look at the funny crow!" And it was my cousin's pet crow that followed him to school, announcing at intervals, "Beat it! The cops!"

Mimicking is also a talent of members of the starling family—esepcially the myna bird and the common English starling, that speckled, short-tailed, garrulous, black bird of our yards and city streets. This lively pest is one of the best of all mimics. A common starling that was raised to adulthood from a naked, potbellied orphan in Texas learned over three-hundred words in two years.

But it is the parrots that have put the verb "to parrot"

into our language. Aristotle named the order Psittaciformes, the *chatterers*. The glib order includes the large green Amazon parrots most people associate with the name, as well as macaws, cockatoos, cockatiels, parakeets, lovebirds, budgerigars, and many more.

A parrot is the ultimate possession for those who long to talk to the animals, not only because parrots mimic, but because these birds just might know what they are saying.

There are arguments on both sides of this old and emotional debate. Psychologist O. H. Mower thought he settled the issue once and for all some years ago. He taught parrots to speak by rewarding them with food when they gave a correct answer to a question. They spoke beautifully, but only what they had been taught, without improvisation, and more out of context than in. Mower concluded that the birds had no knowledge of what they were saying. They were conditioned to mimic to earn a treat, and that was all.

On the other hand, ornithologist T. Luke George of the University of New Mexico, who lived in the same house as Chiquita, an Amazon green parrot, was always greeted by her with an excited "Hello" when he came home. When he departed for work in the morning, Chiquita said "Bye-bye" and flicked her wings as if waving. The words had nothing to do with the time of day. If George came home at noon or at midnight, it was "Hello"; if he went out, it was "Bye-bye."

"I'm convinced Chiquita knows what she is saying," George said. "She never makes a mistake. She has other words in her vocabulary that she shouts from time to time, but never 'hello' or 'bye-bye' unless they are appropriate.

"It seems reasonable," he added. "Birds have greetings and good-night calls. Chiquita simply adopted our words for hers."

Few scientists would have agreed.

And then came Alex. In 1977, psychologist Irene Pepperberg of Purdue University purchased an untaught African gray parrot, a species known for its extraordinary ability to mimic, and began training him with a novel procedure developed by the language expert Dietmar Todt. The method requires two trainers: One acts as the teacher, the other as a rival student. Pepperberg added her own ideas to the training. She taught Alex names for objects he enjoyed playing with or eating. Birds rewarded with food, she discovered, lose interest in learning as soon as they are satiated, so she worked with Alex when he wanted to play, not eat. What is more, Pepperberg and her students used the words they taught Alex in phrases and sentences, talking to him much as they did to each other. They talked about Alex's toys, his food, the color of a box of crackers, and what he was doing. Their idea was to create a naturally talky environment, much like that a child enjoys as it learns to speak. Within a year Alex could correctly name fifty items, including key, pegwood (wooden clothespin), cork, shower, banana, pasta, walnut, wheat, grape, carrot, nut, rock, water, paper, wood, gravel, chair, knee, and back.

A lesson went like this:

The trainer shows the rival (Pepperberg in this case) five sticks.

> "What is it?" she asks.
>
> "Five wood," the rival answers.
>
> "That's right; you're right. Here's five wood." The trainer gives the rival five sticks. The rival plays with them, then holds them out to Alex.
>
> "Okay, Alex, you try. What's this?"
>
> "Wood. Wanna nut."
>
> "No nuts yet; first tell me, how many wood?"
>
> " 'i' wood," Alex replies lazily.
>
> "Talk clearly! How many?"

> " 'i' wood."
>
> "No, not clear enough!" The trainer turns to the rival. "Irene, what's this?"
>
> " 'i' wood."
>
> "Do better. You can talk clearly!"
>
> "Five wood," Pepperberg says distinctly.
>
> "That's right. You take five wood." After going through this routine several more times, Alex finally says "five wood" clearly and perfectly.

The study group worked with Alex four hours a day, the students and professor taking turns as trainer and rival. In this manner Alex learned to count correctly up to seven objects, discern colors and shapes, and speak in phrases and even in sentences. But the question remained—did Alex know what he was saying, or was he just a very talented mimic? Then came this day. A student in the lab was talking about sticks and grapes and colored paper while Alex stared at himself in the mirror. Suddenly the bird spoke.

"What's that?" He kept his eyes on the mirror. "What color?" The student stared at the bird. He was asking an original question. Alex knew many colors, but not gray, and apparently he realized that he did not.

"That's gray," the student replied. "You're a gray parrot." To reinforce the learning, the student repeated the information six or seven times before Alex grew bored and shifted to other interests.

Several days later, Alex saw a piece of gray paper and announced, "Gray paper." Eyeing it more closely, he said, "This is gray paper."

Most people would have been convinced the argument was over. Surely, Alex knew what he was saying.

Pepperberg remains cautious.

"Despite our success in teaching Alex to communicate, we cannot claim that he is a talking bird," she wrote. "Nevertheless, we have uncovered, at the very least, the ability of the gray parrot to use certain English words to influence its immediate environment, affect its trainers' behavior and communicate its needs and wants."

But that is what talking is all about, and my own caution fails me.

"Polly wants a cracker," the parrot in my local pet store screamed repeatedly one day while I was shopping for birdseed. He went on and on without letup.

"Can't you say anything else?" I finally yelled.

"What's it to you?" the bird answered, then counted to ten slowly while sidling along his perch toward me. As he came in close, the pupils of his eyes expanded and contracted in parrot emotion.

"Get lost!" he shrieked in my ear.

I did.

I call that talking.

*"Get lost!"*

# The Horse

Wauwinit, alone in his pasture, stood motionless on the ragged rim of the hill, his neck arched like the Trojan horse. I watched the walnut brown stallion from below, where the sun made stars as it shone through the new tree leaves in the woods beyond the fence. Wauwinit tossed his head. With sudden thunder he swept along the rim of the hill, squealing his defense warning, showing his mane and tail as banners of his speed. He stopped, then trotted to the fence and stared at the stallion in the meadow across the road. Minutes passed, his head drooped, his tail hung low.

"Wau-Wau," called Rose Ann Caris, his handler, who that day was beginning my education in horse talk. Wauwinit turned and focused on her through horizontal pupils. He pawed the ground.

"He's talking to me with that pawing," Rosie told me, "saying he's frustrated, bored. He's begging me to do something." She called his name again. The stallion lifted his tail, thrust his ears up and forward, and galloped toward the gate where we stood. Wauwinit is a Thoroughbred racehorse and a large one. His more than one thousand pounds shook the ground with each hoof fall. I wondered, as he charged toward us, how anyone could communicate with such a powerful and falcon-swift beast.

"Most handlers don't like him," Rosie remarked as Wauwinit pulled to a stop inches from where we stood. "He's cantankerous and unpredictable." I stepped back, although the fence was between us.

"But he's an angel with me," she added.

As if agreeing with both statements, the stallion greeted her with a recognition whinny and a sniff, sweet as an angel; then, cantankerous as the devil, he thrashed his head up and down. Even I could see the gesture was not one of contentment.

"You're in a bad mood," Rosie translated. "What's the matter?" The stallion turned his rump to her in answer.

"Oh, you want some attention? You want your rump scratched? Okay." She rubbed him at the base of the tail. Wauwinit's ears twisted to the side, his upper lip extended and twitched, his head turned slightly to one side, and he sighed. The meaning of the sigh was unmistakable. Wauwinit was as content as a purring cat.

But not for long. In a moment, the stallion lifted his head, and—knees high and toes extended like a dancer—trotted in a circle, and returned to the gate. This time he pawed the ground while sniffing audibly.

"No, I am not going to spoil you that way. You're too fat." The sniff added to the pawing spelled a special kind of attention: food. Wauwinit pawed the ground in my direction and sniffed.

*"Scratch my rump."*

"No," I replied weakly. He was not to be deterred.

"Quit it!" shouted Rosie firmly. Wauwinit understood he was not about to get anything to eat. He turned his rump, but this time he swished his tail noisily in our faces.

"He's miffed," interpreted Rosie. Again I stepped back from the gate. Rosie did the opposite. She opened the gate, approached Wauwinit's shoulder, and draped her arm over his neck. Presently this enormous, testy, and unpredictable animal relaxed and heaved a great sigh of contentment. One thing I understood: This language, so different from that of the birds, cats, and dogs I have known so well, was not going to be an easy one to master.

**PREDATOR MEETS PREY**

The last time I had tried to talk to a horse, he had run with me on a wild ride straight down the steep side of Grand Teton Mountain, ignoring switchbacks and trails and coming to a halt only when his terrifying descent had taken us from the treeless alpine tundra to well below timberline. He hadn't listened to me; I hadn't understood him.

To converse with a horse, one has to have an even more precise knowledge of the nature of the animal than is needed to communicate with bird, dog, or cat, if for no other reason than that mistakes can be, if not fatal, certainly injurious. My friend Bob Milton, once a jockey, now runs an automobile garage because of a misunderstanding with his horse which resulted in a fall that badly injured his legs.

Horse talk is difficult for us to interpret because the animal is a prey species, which we are not, and many of its communications are long-distance telecasts designed for the wide-open spaces of grassland steppes and prairies. A horse's face is not as expressive as a cat's or dog's, which are predators like ourselves. But they are social mammals, and, like all social mammals, they speak of rank, affection, and of their intention to defend themselves against those who irritate, affront, or frighten them. To the extent that our social lives also express who we are, whom we like, and what upsets us, we can learn to talk with horses.

The relationship between human and horse is unique—a prey species permitting itself to be handled by a predator who is smaller and more frail than itself. Because of this, communications between the two are iffy and, at times, downright mystical. Writes Beth Ferris, award-winning screenplay writer and novelist, of her communications with her spicy mare, Sunny:

"I am amazed by the sensitivity of my horse, always and forever. She can sense the smallest mood change in me, almost, I feel, as though she were reading my mind before I speak or move. I have several times experienced this uncanny feeling with her. But I suppose a prey species must be sensitive to the point of unfathomable.

"It is of crucial importance to understand that the horse

is a prey animal that lives out in the open, and runs when it is frightened. A horse has to be 'won.' It has to overcome its instincts so that we may ride it." Beth admits that her relationship with her horse is a kind of truce, and only a partial truce at that.

"She doesn't love me, she may recognize me in some way, as a friend. I think a lot of people fool themselves into thinking that the horse has real affection for them. I don't believe it. It is more like common ground, a meeting place, where we agree to accept each other's horseness and humanness for a short time and communicate that to each other."

## HORSE-TALK ROOTS

The horse, *Equus caballus,* is a domestic animal, but like the cat, it can live wild without us, for its behavior has been so little changed by breeding that each young horse must be "broken" to our needs. Each filly and colt, despite early handling, fights the bridle, harness, saddle, and rider until, slowly and patiently, a more dominant individual quiets its resistance.

The horse did not move in with us by its own consent as did the cat, which enjoyed the rodent-rich ecological niche humans provided, nor had it evolved to please like the dog. It may even be that horses were captured first not to help man actively, but to protect his crops from their grazing and trampling or to keep meat on the hoof handy.

Even now, four thousand or perhaps as many as six thousand years after their domestication, that is the economic role of horses among some of the Far East peoples of the Soviet Union. The historical threads of the domestication are tangled. Wild horses flourished in steppes, forests, deserts, and tundras. At the time of its domestication there were various breeds of *Equus caballus*—the tarpan, for example, and Przewalski's horse, the only wild breed that still exists. Which breed was first bred by humans, and whether it was later mixed with other wild breeds, was not recorded, and the progress of horse domestication, like that of the cat and dog, will probably never be traced.

But mankind must soon have seen in the big animal swifter legs for the hunter and stronger beasts of burden than women and dogs. Around 2000 B.C. the Mediterranean people harnessed the horse to lightweight, spoke-wheeled war chariots. That advantage had dramatic effects. Military conquest spread the knowledge of herding and training, and horse culture moved into Europe. It revolutionized not only war but travel and farm work. The revolution was responsible for the almost total disappearance of wild populations. Wild herds were competition for the land and water needed for domestic stock, and there were profits to be made in capturing and selling these valuable beasts.

More is known about the evolution of horses than of any other mammal, and the story revealed in fossils is one of extraordinary change.

The earliest identifiable ancestor of the horse was a tiny—ten to twenty inches high—forest dweller called *Eohippus*, which browsed on soft weeds and leaves, had spreading toes to keep it "afloat" on squishy ground, and probably led the shy and solitary life of small forest prey today. That was fifty million years ago in what was then the subtropical forests that covered nearly all of North America.

From that time on, worldwide climate gradually became drier and colder, hardy grasses replaced tender growth, broad plains spread where forest had once offered protection, and living out in the open became a necessity. Everything about the modern horse is an adaptation to prairie life.

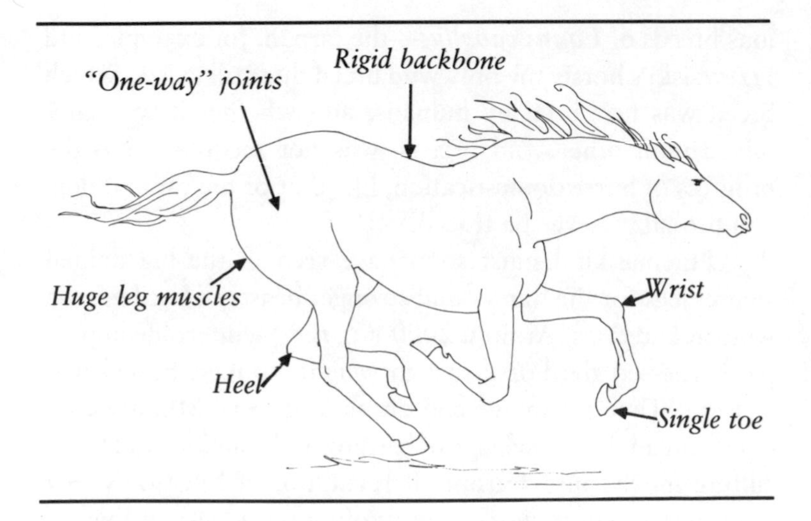


"They run," Beth said. They run today because the prairies where they evolved offered no place to stand and hide. Horses' legs grew longer, until a racehorse's running stride is now four times the length of the body. It pushes off like a pole-valuter on one-toed feet, armed against wear by a thick toenail, the hoof. People have trouble drawing horses—what seems to be a "knee" on the foreleg is really the wrist; most of a horse's "leg" is foot.

The horse is in every way a running machine. The limb joints move in one direction only, back and forth, to save energy. The backbone hardly bends at all; effort is concentrated in the enormous leg muscles, not wasted in a bending back. As the horse developed speed, its eyes moved higher on the head so that it could see a potential predator even with its head buried to graze on lush grass.

A horse's eyes are very large, larger compared with its size than a whale's eyes. The horizontal pupil scans the horizontal world of the open grasslands almost equally well to the side, the rear, and the front. The visual field is an extraordinary 215 degrees, a veritable cyclorama. A blind spot lies to the rear in a small area, but with a slight turn of the head and neck that is taken care of. Horses are farsighted; focus is poor close up but excellent for distant objects. A horse can see, for example, a stallion silhouetted against the horizon five miles away. The eyes focus binocularly in front so that the horse sees where it is placing its feet in three dimensions, and it can, at the same time, see monocularly to both sides without depth perception but with precision.

As the horse adapted to life in the open, its behavior also changed. Where it was better to be a loner in the woods and hide from sight, it became better to stick together out in the open. That way there were more eyes and ears to detect enemies.

Sticking together in a group required a communication system to keep order in the group and to warn one another of danger. Horse talk depends a good deal on big postures and dramatic action visible from afar. A stallion like Wauwinit strikes his *boss* pose on a high point of land. He arches his neck and lifts his flowing tail. If pose alone does not discourage an intruder, he rears up on his hind feet and paws the air—a big gesture for a big country.

The horse alarm is also vivid—to give it, the horse runs. Herd members, even with heads down grazing, see the runner on the cyclorama of their vision and run, too. So deep is the response to the running alarm that a horse will follow any member of the herd that runs whether the runner is leader or a lower-ranking individual. This response explains why we are able to race horses and why inexperienced riders are usually confined to a ring.

The ring is all the more necessary because horses are easily alarmed and have a superfast reaction time. The horse

*The boss pose*

nearest to "danger"—it can be a car's backfire or a startled rabbit—pivots and flees. The others follow on the instant. The flight response elicits more alarm, and that sparks more, until they are going faster and faster in a stampede.

Horses are not territorial; they do not stay in one area and fight for it, for which we have reason to be thankful. They are nomads that, during the Ice Age, found their way to Europe and Asia across the temporary land bridge between Alaska and Siberia. They survived the rigors of the Ice Age in North America only to become extinct in their homeland as the last of the glaciers melted away. All horses now are descended from those that migrated to Eurasia.

Horses were reintroduced to the continent by Columbus on his second voyage, when he transported seed stock to

Santo Domingo. There, and on other stock islands, horses were allowed to breed freely, then rounded up to aid in exploration, conquest, and settlement. Now transformed into a domestic animal, *Equus caballus* was back home.

The horses of Columbus, Cortez, and Pizarro were intelligent Spanish horses bred in Andalusia and Seville during the sixteenth century. These tough, small horses had no peers for endurance and speed.

On the backs of these durable little horses the Spanish pressed north and west into the territory of the Plains Indians who followed the buffalo. It did not take these Indians long to observe how the Spanish trained and used their horses. The Apaches, Utes, and Navajos stole Spanish horses or captured escapees, applied their acquired knowledge, and by the 1800s had become among the most skilled horsemen in the world. They rode without bridle or saddle, clinging to the mane braided to either side like reins, and used the animal as a shield in battle by slipping over the side to protect themselves behind the horse's flank. Using northern European stock as well as Spanish horses, they developed their own breeds, the pinto and palomino.

But in the end, Europeans wiped out the buffaloes and most of the tough little horses, and thereby most of the Indians. The survivors managed to retreat into the desert, taking with them their own prized breeds, as well as a few of the last Spanish horses.

## WILD TALK

It is from these stocks, escaped from captivity during the early 1800s, that we have learned most of what we know about horse communication in the wild. Free again on the grassy savannas in remote valleys, the escaped horses took up a wild life as if they had never left their homeland. They were admirably suited to the strenuous adjustment. So successful were they that by the mid-1800s there were an estimated four million wild horses running the Great Plains with the last few buffaloes and pronghorn antelopes. Called mustangs from a Spanish colloquialism *musta,* "one who owns horse stock," the wild horses today have been reduced to six thousand due to human expansion and to the very nature of horses: their urge to herd and to be herded.

Horses' "natural" leader is the boss stallion who, when danger threatens, takes a protective position behind his herd and directs them out of danger. From this position the stallion whinnies and squeals, pounds his hooves, and blows loudly through his nostrils, orchestrating his band and "rounding them up" for escape.

This behavior, which once served survival, is what nearly did in the last wild horses in America today. Wranglers, taking the role and position of the stallion, herded wild horses to where they wanted them to be, often into a dead-end canyon where they were slaughtered for dog food.

Thanks to two dedicated women, Hope Ryden and Velma Johnson (whose crusade on behalf of wild horses earned her the nickname Wild Horse Annie), America's last wild-horse herds are being managed and protected on three refuges where their behavior can be studied.

Given their freedom, horses live in bands containing females, their daughters of any age, young male offspring, and a single stallion. Other males may challenge and fight the stallion during the breeding season, and if he is conquered, the new stallion takes over the harem and drives the old horse out. Females come into heat during the early spring and, if not bred, repeat the estrous period every three weeks until fall. Gestation is about 330 days, so that the one or occasionally two foals are born the following spring when the food supply for mother and young is ample. The young reach adult size in two years and, like many mammals, are sexually mature somewhat earlier. Both sexes may live more than twenty years; a fifty-two-year-old, the oldest verified age, died in 1970 in Australia. The horse weighs from eight hundred pounds to more than a ton. The heaviest on record is an Argentine horse named Brooklyn Supreme who weighed thirty-two hundred pounds and stood six feet six inches tall at the shoulder. The Paul Bunyan country of America developed the strongest horses. A pair of draft horses hauled a load of lumber weighing 59.2 tons in Michigan in 1893, for the world record.

### APPROVING REMARKS

For all their size and strength, and at any age, horses are like high-school kids: Their most serious concern is being accepted by their group and following their leader—even down the wrong road or into the slaughterhouse.

Rosie insists their psychological age is even younger than a teenager's. "Horses are simply big children," she told me. "They have the mentality and needs of a two-year-old kid. And they can be spoiled. I spoil Wauwinit." Tanrackin Farm, where Rosie works, is home to Thoroughbred mares and stallions living under the necessarily unnatural conditions of domestication. Stallions are too vigorous and dangerous to be confined in fenced pastures with other horses, and so Wauwinit is alone in his field day in and day out. He is thus deprived of the approval of his herd, of either leading or following, and of the companionship with pigs or chickens that even a barnyard provides. Rosie's duty at the extensive racehorse-breeding farm is to be a friend and handler to Wauwinit and several other valuable stallions and mares. A central gesture of that friendship is the one that had so quickly brought the stallion out of his grumpy mood—laying an arm over his shoulder.

"The arm over the neck is horse talk," she explained. "Wild horses, and even domestic horses, lay their heads over the necks of their companions to express friendship. I just told Wauwinit I cared for him."

*Friendship, person to horse*

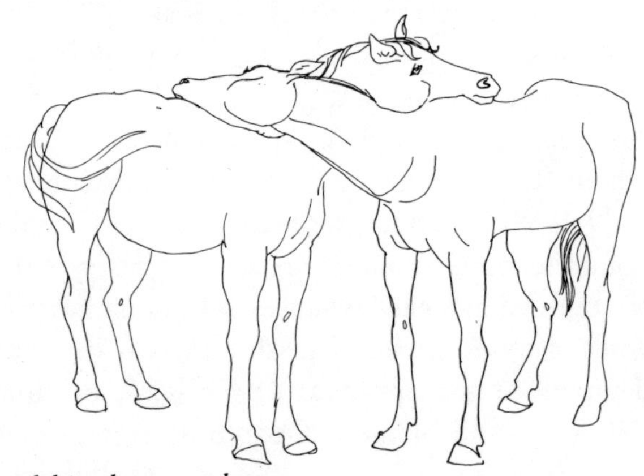

*Friendship, horse to horse*

Mrs. Waller, owner of Tanrackin Farm, used this need horses have for approval to bring a valuable horse out of the dumps and on to a championship in the famous annual horse show at Madison Square Garden in New York City. Queenie, a Pollyanna pony of Welsh and Arab descent, had been a strutting princess among her peers at a farm in England when she was bought by a wealthy American oilman and shipped to his ranch in the Middle West. There she was put out among his other horses and more or less ignored, not physically but conversationally. Mrs. Waller had seen Queenie in England and remembered her as a lively, proud thing. When she came up for sale, she purchased her.

The little pony was not the animal she expected. She was uninspired, dull, and walked dejectedly. Word was sent out to all the handlers, "Every time you pass Queenie, tell her she's beautiful. Tell her she's a queen."

After a year of this praise and attention, Queenie was prancing again, her head up, her eyes bright. She was once more a sassy little queen.

This need to be "one of the gang" cannot be genetically manipulated. The herd mentality is the result of millions of years of evolution and is not to be turned back in a mere few thousand years of mankind's selective breeding. To convince a horse that a human is a fit companion, the human must abide by the rules of the herd, both as friend and boss.

## BOSSINESS AND SNOBBISHNESS

Observers of wild herds report that the rules that hold for a herd on the savannas hold for domestic herds. There is a boss or leader, a second horse, a third, and so on down the line to the bottom. This order is known among horsemen not as the pecking order but as the *kicking order*. Rosie calls it the *kick-and-bite order,* and that is both more descriptive and more accurate. If a leader is feeling irritable, he bites and kicks number two, who passes it on to number three, and so on to the last, who, with no horse to bite, often takes it out on the donkey some horse owners stable with their horses for company.

On a horse farm this hierarchy breaks up into two other hierarchies: the feeding order and the social order. One horse may beat all the others consistently to the food trough, but he or she is not necessarily the lead horse in other matters. Another, more self-assured horse makes all the social decisions, such as where to go and when—or even whom—to harass. The feeding hierarchy does not exist in wild horses, which, of course, are free to graze full-time.

Among a group of domestic horses, the dominant horse is not necessarily the biggest, strongest, or the most experienced. Age has some influence on rank in that horses under five usually defer to their elders, but temperament—or, rather, temper—is the stuff that horse leaders are made of. In a stable up the road from my home the dominant horse is a small but aggressive mare. She is called the lead mare, and a lead mare she is. If she hears a car engine start up suddenly, she squeals, puts back her ears, and runs. Instantly the other members of the herd squeal, put back their ears, and run. When she stops, the other horses stop. If any horse comes too close to her, she bites its rump, and it bites the rump of the next-ranking horse below it, and on the bite goes to the bottom of the line.

In wild herds, males are dominant in leading and in organizing defense. A wild stallion who lived with his herd in the rocky crags of the Red Desert, Wyoming, saw above him an approaching helicopter that he recognized all too late to be an enemy. He herded his group for escape and summoned two other stallions to defend against the novel predator from the skies. The boss stallion rose on his hind legs and flailed; the other two rose on their hind legs and flailed while the craft hovered and the predator shot every one of them.

Wild lead mares, however, make day-to-day decisions for the group: to head for water, to lead the foals to new pastures.

The personality on which leadership in domestic horses is based is inherited. The offspring of feisty lead mares are often feisty themselves. But there is also a cultural factor. Upper-class mothers tend to cluster with their social equals—other high-ranking individuals. The offspring learn their p's and q's among their "own kind" and also make the right "connections" that will serve them in adulthood. In domestic herds, unless people split up these snobbish social groups or separate mares and foals, the offspring of the upper-class mares move up the social ladder quickly by having their mothers around to push them, just as they do in the wild.

## SMOOTH TALK

The need for space is another herd law, and apparently a serious one, for intrusions onto personal space are punished. Social life in a horse herd is extremely physical. When strangers are put together, they herd, but they also bump, shove, and push to gain their private space. The punishment for not yielding is a kick, or bite, or both. Once a herd has settled in and each member knows its place, the high-ranking horse need not fight for its space. It simply pins back its ears, swings its hind end toward the offender, and space becomes available.

A trainer cannot bite or pin back the ears to tell a horse who is boss, but he or she can kick or swat with a stick—

and some do. Some do not. Rosie is a believer in nonviolence, although she is not above striking a horse that has exposed its teeth and is rearing to strike. "Then, sure, I've got to use horse talk and strike back or I'll be killed. But, for the whole route, I don't like it."

Like Mrs. Waller, who talked Queenie out of the dumps, Rosie soothes potential aggression with her voice when she can. I tried out her advice on Wauwinit. After Rosie had gone into the stable, I ambled back toward the gate and slowly held out my hand for Wauwinit to sniff. Fast movements and sudden approaches frighten a prey species; sniffing is a way a horse familiarizes itself with a new person. Wauwinit sniffed; I talked, telling him how handsome he was, about the meadow, the trees, Rosie, and anything else that came to mind. I noticed Wauwinit's strange pupils taking me in. His eyes closed slightly, as they had when Rosie put her arm over his neck. For the first time in my life, I felt I was communicating with a horse.

But man-horse confrontations call upon the use of horse talk as well as human talk. Several years ago I took notes on a most impressive confrontation between Gilbert Strang, a

dairy farmer of Ypsilanti, Michigan, and his standardbred stallion, Jube, when he went into the field to bring the horse into the barn.

### A MAN-HORSE CONFRONTATION

"Yo, Jube!" The horse heard his name, turned his head, and swiveled his ears toward Gib. Some days he would walk calmly to meet Gib; that morning he did not. He ran full tilt to the horizon, where he arched his neck and pricked his ears forward. True to his sex, he was a masterpiece of aggression and defense, as ready to strike as he was to run. The wind blew his mane and forelock, making him appear larger than he really was, enough perhaps to scare off a challenging stallion—and certainly me.

He gave Gib, who was now in the field walking toward him, the threat stare designed to intimidate the enemy. Using horse talk, Gib stared back. Jube's threat having failed, he took the next step in a horse-to-horse confrontation. He squealed the high-pitched outcry that is a warning that the horse is prepared to fight.

Gib walked on, lowering his head aggressively. Jube lowered his head aggressively, switched his tail, and vigorously pawed the ground in annoyance. I was half the field away, but this bold semaphore was perfectly visible. Gib answered Jube's "miffed" signal by shouting in English. Jube knows about eight English commands, including "Git in here," but this time he ignored the words. Gib, using big grassland gestures, held up his hand and walked toward Jube's shoulder. That is where Rosie also had aimed her approach to Wauwinit, and for good reason. The shoulder is a horse's nonaggression zone, where a touch is least likely to elicit a kick or bite.

So far, neither had the advantage, and Jube knew it. Gib did not run away as a lesser-ranking horse would have done, so Jube asserted himself further by drawing his head toward his chest and holding it there. This gesture is a statement of presence, a sort of king-of-the-mountain—or, rather, grassland—challenge. Gib contradicted that opinion by walking steadily forward, straight toward Jube. He added a touch of majesty to himself by holding his elbows out from his body to accentuate his size.

Jube thrashed his head up and down, a sign of some apprehension as well as a rebuke, then warned Gib to stay back by threatening with lifted forehooves. He lifted them higher still, then pounded them back to the ground with a noisy thud. Gib still came on.

Rearing, Jube shot his front legs stiffly forward, a threat to strike that is to be taken very seriously.

Gib was almost close enough to grab the halter when Jube lowered his neck and extended his head aggressively again. Gib extended his neck, too. Jube's lip went up, exposing his teeth in a threatening-to-bite look. At that, I heard Gib talking in a deep voice, talking, talking, soothingly, calmly. Jube hesitated and looked at him curiously, possibly because in the horse world, a noisemaker is not a predator. A predator comes silently. There is no sound when the wolf hunts or the cat strikes, no chatter when the owl flies. Horses may not be capable of learning many human words, but talking softly and steadily reassures them. Jube, however, recovered quickly from his confusion and reared again.

*A one-legged threat to strike*

*A two-legged strike*

**THE FECAL CEREMONY**

Gib reached for the halter. Jube tossed his head, and Gib missed. A predator, no; but a challenger, yes. And the stallion was not ready to give in. He initiated the battle ritual of the stallions—the *fecal ceremony*. He deposited a steamy pile.

In the wilds two stallions sparring for dominance, as Gib and Jube were doing, trot up to each other after dropping their feces and learn one another's identity by sniffing noses, then noisily exhaling. They simultaneously have erections visible to one another—possibly a statement of arousal, albeit an arousal to fight—then they sniff each other in ceremonial sequence: first the neck, then the flank where the hind leg joins the body, then the withers at the top of the shoulder, then the back, the genitals, and finally the rump.

That ritual completed, they place heads together and shove. After this preliminary muzzle punching, they break apart, sniff each other's piles, and urinate on them, putting their own signature over that of their rival. If by now the defending stallion has not turned the other away, he gives the high-pitched outcry of furious energy known as the squeal—"*E-he-eeeeee!*"

The rival answers by defending himself, rearing up on his hind legs, exposing his teeth, and flailing his hooves.

If neither gives in, they fight with thundering violence, bumping and shoving with heads, necks, and shoulders, aiming bites at hindquarters, and, finally, turning the rear, where the power properly lies, to kick. Kicking combatants aim not directly at the other but off to one side. Violent and frightening as a horse fight looks, it is designed not to injure. After a brief onslaught and if round one has been a stalemate, the two stallions dash apart in preparation for the next round, announced by the dropping of a fresh fecal pile. Horses eat great quantities to get the needed nutrition from grass, and so they have manure to spare. These piles, hot and steamy, apparently reek of wrathful odors, or so it would seem from the fury that is now unleashed. The battle is won only when one stallion, rising on his hind legs, tucking his rear under, and flailing like the wild stallion attacking the helicopter, rises to such a height, sustains his attack for so long, and strikes with such ferocity that the other backs off, turns, and gallops away.

Although horses are not killed in these encounters, humans sometimes are. Gib had to assert his dominance before the confrontation escalated to an attack. As Jube crowded, shoving with his neck, Gib grabbed the halter. Jube turned his rump to tell Gib off, but before he could kick, Gib gave a well-aimed kick toward his belly. Kicks Jube understood. He backed up and stood quietly. Gib was at the top of the kicking order, and Jube took his position in the man-beast hierarchy. He let the "lead stallion" guide him quietly across the field, down the lane, and into the barn.

## TALK AMONG FRIENDS

Friendships are a very important part of the social life of a herd, and most horses have a "preferred friend." Best friends are, generally speaking, two horses closely matched in rank, size, and disposition. Horse friends are often inseparable. The deepest bond occurs between a mare and her female offspring, a relationship that can go on for years and years and that comes to misery for both when the animals are separated. Many horses balk at being separated from their best friend, which, horse owners know, had best be kept in the adjoining stable.

A person can also be a horse's friend, which is what

Diane Guillano is to Audacity, her gelding (an altered male). Each morning they communicate in horse talk much as Audacity does with his best horse friend, Stone Wall, who is stalled beside him at the Beech Hill Farm riding stable not far from my home.

Before dawn one day when, intimidated by my stallion education, I had arrived to observe tamer specimens, Audacity came out of the trance of sleep and pricked up his ears. He had heard a distant, familiar sound. His eyes softened as the lower lids moved upward. I did not hear any car for at least a minute, then I heard two that sounded the same to me. One pulled into the riding school parking zone, and presently the stable door opened. Audacity stamped with one foot and immediately gave a special whinny known at the stable as the "Here-comes-Diane" whinny. As she strode down the ramp into the stable, Audacity stuck his head over the stall door. He saw her and whinnied several times.

"Hello, yourself," she called. "How are you?" Audacity stamped his foot again. None of the other horses answered. Diane was speaking to Audacity.

She opened his stall door and stepped in. As the dawn light came through his window, the gelding's spotted gray coat took on a silvery sheen. He moved his weight from one foot to the other, arched his neck, pulled his head toward his chest, and arched the fleshy part of the tail in a moderate version of the stallion's proud "I am me. See me."

Diane was interested in obtaining a bucket on the other side of the stall. Audacity thrashed his head up and down like a pump and flourished the long hair of mane and forelock that tumbled between his eyes. But Diane was still preoccupied and not taking enough notice of Audacity to suit him. He stepped forward and sniffed her audibly. He put his head against her. He whinnied, but in a slightly different tone than the greeting whinny. It sounded like a question, "See me?"

Diane was too intent on reaching the bucket to pay attention to him. She needed to get around him, but Audacity practically filled the stall. She told him to move over by putting her hand on his neck and pushing: "Space, please." He moved, but not far enough. Diane pushed him again, this time with more strength and authority.

Audacity turned his rump to her: "I'm getting mad."

"Well, you'd better not," she answered in English, "because I'm going to saddle and ride you this morning." Her voice was loud and steady, but it was slightly high-pitched, as if she were speaking to a child.

Audacity was not ready to be pushed around. Even the best of horse friends exchange harsh words occasionally, and this was one of those times that tries friendship. His ears went back in anger. Diane made herself look bigger and more dominant by taking a wide stance and putting her hands on her hips to say in action, "I'm boss." The gelding understood the stance and tone of voice, and backed up. He opened his eyes wide and held his head to one side to say, "I am submissive."

"Glad to hear it," she said, hugging him and stroking his neck. Audacity sighed. Their friendship was undamaged.

In the Beech Hill Farm stable, owner John Schrader knows every friendship among his horses. He tries to send them out into the ring and on the trail with friends together and enemies apart. He also puts friends in stalls next to each other because it makes for a more peaceful stable.

"The stallion can be a problem," he admitted. "That's why I put him in the end stall with no one on one side and the lead mare on the other. With her there he is quiet and peaceful." Most horsemen do not put stallions next to mares, but in the case of this stallion and this mare it works well.

"You have to know the personality of each horse. I have seventy-five here, both school horses and boarders, and it's important to stall them so that they will not be threatened. They like to feel like a herd, but," he added, referring to the fact that herds include enemies as well as friends, "I can't let them act like one."

## GROUP LESSONS

The herd is not only a mix of enemies and friends that shape a horse's life but also a school of learning.

A recalcitrant filly, Miss Tips, who lived on a farm near our summer home in Pennsylvania, lorded it over the goats, chased the piglets into a corner, and bit the calf when she could. Furthermore, she would not be led by man or beast. The day she kicked her young mistress, the girl's father promptly took Miss Tips down the road to horse school—a herd that spent most of its time in the fields. Miss Tips was pushed through the gate and left alone with them.

She immediately asserted herself. She shoved, bit, and kicked for three days.

Then the tables were turned. The lead mare kicked her, hard and painfully. The other mares and fillies nipped her rump and neck, and a gelding ran her into the corner of the field where she reared in defense, flailing and squealing. After several weeks in this school of hard knocks, Miss Tips had been told, and had accepted, her position in the social registry—second from the bottom.

By the time she came home in the fall, young Miss Tips had learned her manners. She no longer terrorized the goats and bit the piglets but took her place among the farm animals and soon she was walking on the training lead called the *longe*. The herd, not people, had socialized her.

The herd's teaching gives a young horse confidence. The handlers at Beech Hill Farm regularly let an old horse lead the new and young ones onto trailers, over jumps, and around the arena. On the Tanrackin Farm, experienced older horses called *lead ponies* help introduce the two-year-olds to racing. They take them around the track and through the starting gates to teach them what they must do.

The dispositions of these teacher animals is very important, for they are lead animals, and the attitude of the leader trickles down to those that are led. The horse that sets the pace might better be a nice one than a bully. If the lead animal is a troublemaker, herd mentality works in such a way that there will be troublemakers from the top of the hierarchy to the bottom.

"If you put a nice horse in a field with a rotten horse," Rosie told me one day, "you'll get two rotten horses, not two nice horses. Somehow horses pick up bad habits faster than good ones. I show my animals how to be good by putting them with good animals."

A bad horse is one that bucks, bolts, shies, or *cribs,* which means it chews the stable. An even worse habit is called *wind sucking*. The horse takes the edge of a door in its teeth and sucks, gagging and filling its stomach with air. Wind suckers do not eat well and weave around the stall banging their shoulders and heads. Such repetitive, self-damaging behaviors can occur in captive animals whose lives are not in keeping with their natural urges. Or, as Rosie put it, "Cribbers and wind suckers are telling us they are bored." Once started, these habits are hard to break. To help reduce the problem, veterinarians recommend companionship, along with regular exercise and a diet high in fiber. Returning the horse to the pasture where it can expend energy foraging often eliminates the habit, particularly if it is among its fellows.

**BRIEF ENCOUNTERS**

Breeders agree that the more a horse is educated, both by good human friends and through pleasant equine companionship, the more likely it is to become and remain a tractable animal. The earliest social environment of a horse is, of course, easiest to control if the owner is also the breeder.

The sexual communication between domestic horses is, almost without exception, very carefully supervised. Horses are too valuable to be permitted to mate with the horse of their choice for two reasons: genetic and physical. Breeders who have invested a good deal of money and effort in their stock wish to keep or improve a genetic standard, and courtship among horses can be vigorous to the point of injury. Courtship is simply eliminated. When the mare is ready, the stallion is taken to her at the precise moment of fertility.

On the Tanrackin Farm, where the breeding program is a foremost endeavor, the breeders discern when a mare is ready by leading her to the *teasing board,* a large wooden obstacle, behind which stands a stallion. The stallion sniffs and licks her vulva. If the mare kicks or the stallion turns away, she is not ready. In addition to this test, a veterinarian does an exam, but the horses make the final decision. When the mare lifts her tail to the stallion behind the teasing board, moves it to one side, and urinates, she is ready. The pair are led into a protected arena for breeding. She is held on a lead rope, and he is held on a stronger lead called a *chain shank.*

"It takes a very short time," said Mary Ann Kamph, manager of the mares and foals at Tanrackin. "The timing is so right; he sniffs, mounts, and ejaculates. As soon as he dismounts, he is led away."

On a wild-horse range in New Mexico a stallion keeps track of the progress of his mares' estrous periods by frequently smelling their urine and their vaginal regions. As one comes into heat, he begins to court her. He whinnies when she is near, keeps his eye on her, and approaches her from time to time to rub his neck against hers. If she is not ready, she will back off, kick, and run. He doesn't follow. She can kick hard.

As her period of fertility nears, the wild stallion approaches her more often to smell her vagina, whinnying an inquiring cry. It is the softer whinny Audacity gave when Diane ignored him. The question in this case is, "Are you ready?" If she does not run, he prances toward her. A guttural nicker pulsates through his head from his throat, a nicker rarely heard except when a stallion is testing his mare. It is a sort of "I am ready. Are you?" If the mare does not squeal in outrage and thrust her front feet at him but sniffs the air and looks at him with soft eyes, the stallion comes up to her. She permits him to touch her.

Head outstretched, he smells her head, flank, groin, and genital areas. He nips her thigh playfully. She stands still. With his tongue he caresses her neck and her forelegs and hind legs. His penis lengthens and swells. She turns her head and looks at him, lining herself up and encouraging him. She lifts her tail and urinates. She is ready. With that the erection becomes complete and enormous. The stallion rears up and mounts. He places his forelegs firmly on either side of the mare's body, nestling his neck against hers. She moves her tail to the side, inviting entry.

As the stallion penetrates, he grabs in his teeth her mane, neck, or shoulder—anything he can hold on to while he thrusts, penetrates, and ejaculates. The surprising speed—about a minute, according to Mary Ann—is typical of copulation among herd animals, for no prey species can risk lingering long at anything that renders it helpless against predators.

The stallion dismounts, and within sixty seconds the penis has retracted into the sheath that normally covers it. He walks off.

## AN EARLY START

The eleven-month gestation period is a time of watchfulness, excellent feed, and much grooming of the brood mare. March and April Mary Ann takes up her vigil in a room in the barn. She naps during the day, for most of the foals are dropped at night, and she is ready for each one.

"I start talking to them the moment they are born," she said, "getting them used to humans." She may help by giving a little pull, and she does aid the mother by taking the membrane off the foal's nose and head.

"Then I leave the two alone for about twelve hours." Being touched and talked to at birth, she explained, is enough for now. Her further presence might interfere with this crucial period of bonding between mare and foal.

On May 1 a colt was born that Mary Ann immediately took a shine to. He was delivered speedily and reacted to her help by getting to his feet, wobbling, breaking at the knees, and finally taking a wide stance and holding it.

Then she left him to enjoy without distraction his mother's first communications to her new baby—her caressing tongue as she cleaned him and then a bunt of encouragement as the foal smelled the mare's milk and searched for the source.

On the second day of the foal's life Mary Ann came into the stall to put a halter on him and was kicked, butted, and bitten. "He's a tough little fellow," she said warmly, "so I called him Spike. That's his barn name."

From that day on Spike was handled, petted, and talked to endlessly by Mary Ann and her niece Ann, who spend the spring months acquainting the foals with the sound of voices, the touch of hands, and the feel of the halter and lead rope.

When the spring grass pushed up and covered the rolling hills, Spike, his mother, and the seven other foals born that year were led out with their mothers into the field to graze each day. When Mary Ann got Spike up in the morning, he would bite her and jump away from her. One morning he slapped her leg with a front foot.

"Very naughty," she said.

I wondered how that remark could possibly be understood by a horse.

"I say it in English," Mary Ann told me. "Then I scold him in horse talk by pushing against him like horses scold each other."

### LOVE A BIG BABY

On a golden September day I met Spike for the first time. He was on his way to the meadow with his mother, frolicking and looking in all directions. Two yearlings that had been tacked up with bridle and saddled for the first time that morning passed by. Like the wild things they are, they were fighting the bit and squealing. The handlers dominated with strongly stated words and movements. Even as I watched, the yearlings lowered their heads and walked quietly down the lane.

They were on their way to be broken—ridden for the first time. I remembered a remark Beth had made. A horse, like a child, has to be slowly trained. They take one step forward, one-half step back. She said that training a horse requires great patience and persistence; they are trained only by repetition, trust, and human understanding that they are really wild animals.

Around four in the afternoon Spike's mother was led back to the broodmare barn with him at her heels.

Mary Ann approached the deep brown colt. Spike trotted up to her, then frisked in a little circle before her, as he did to his mother to solicit her attention and care. She dropped to her knees and held out her hand as if it were the nose of a horse extended for a recognition sniff. Spike *snapped*—that is, he thrust his head under his forelegs, moved his jaw vertically, and, drawing back the corners of his mouth, clattered his tongue softly. I had never seen a horse make that gesture before, nor had I ever heard that odd clatter. Spike was acknowledging that Mary Ann is a leader and that he is submissive.

The signal elicits affection among horses, and Mary Ann gave Spike the care he had asked for by pressing his head to her shoulder. This, like draping an arm over a horse's neck, is a human version of the horse's pose of affection in which the head rests on the neck or shoulder of a friend.

The feeling that emanates from Mary Ann when she is with a colt is one of tolerant pleasure, like a grandmother with a grandchild. She inspires confidence in these sensitive youngsters by frequently touching the nose, head, neck, and shoulder—the spots where horse touches horse. She shows confidence in her bold movements and firm voice, which are akin to the kick and squeal of horses in intensity if not in tone. The foals acknowledge her solicitude and firmness by frisking for her. The early handling of a foal can give it a lifetime of good attitudes toward people. Bad attitudes, equally possible through unsympathetic handling, are almost impossible to erase after a horse is two years old.

## FINER MEANINGS

The tension between predator and prey, man and horse, never entirely lets up no matter how well or how early in its life a horse has been handled. At least the horse is honest: It tells

THE HORSE'S FACIAL EXPRESSIONS
"All's well."
"There's something alarming behind me."
"There's something alarming down there."
There's something interesting down there."
"I'm annoyed."
"I'll bite."
"I'm boss."
"You're boss."
"I like your touch."
"I smell . . ."
". . . something"
". . . sexy!"

the handler how it feels before acting upon its feelings.

"Watch the eyes," is Rosie's advice. "They tell whether the horse is feeling kindly, mad, sick, or is not to be trusted."

Wauwinit's eyes roll, showing the whites, when he is angry with her or with one of the stallions in the barn.

In the "I-am-not-to-be-trusted" look the pupils enlarge in his wide-open eyes and the eyeballs shift in the direction of the action to come. I thought that a valuable piece of information. If a horse is going to rear, its eyes flick briefly upward. They shift swiftly from side to side if it is going to kick.

When in a good mood and when he is feeling comfortable, Wauwinit's eyes are open and round. The bottom lid moves up when he is feeling friendly toward Rosie. The eyes are dull—the lights do not flash in them—when he is not feeling well.

Most of us watch the ears of a horse for news of its mood, not only because they are less subtle signalers than the eyes, but because we are usually riding the horse's back and can see only the ears.

Ears held in a normal position, with the openings facing to the side, mean "All is well," the horse is at peace with its environment. Riders who note this relaxed expression look off and enjoy the scenery.

Pricked, the ears announce a sound on the wind that the horse feels it had best pay attention to. Audacity pricked his ears forward to announce a fly coming through the stall window. That it was a fly and not a bird or mouse was confirmed by his tail. He swished it in anticipation of the pest's bite.

An ice patch got me into a conversation with a horse that began with her ears. We were winding through the gray forest on a cold November day when Honey stopped dead.

Her first statement was to lay back her ears to say she was afraid of something. I clicked "Giddyup," slapped her neck, and urged her on. Her next comment was to toss her head up and down—irritation. I looked down. A slick of ice was on the trail, and she would not take another step. Even a snap of the whip did not start her off. Thinking back over all the advice I had been given by riders for situations like this, I came up with Beth Ferris's words: "Think like a horse."

I glanced down at the ice again. It gleamed ominously. I would be frightened if I were a horse. Feeling sympathetic, I looked for a way around the treacherous patch, and, as I did so, my annoyance changed to concern.

"Go over there," I said, pushing with a knee and pointing. (I am no horsewoman. I point things out for the horse, let it argue with me about which path to take, and even grab the saddle in awkward situations—something my Maine camp riding teacher told me never to do.) Honey apparently sensed my solution through the way I was leaning. Her ears began to lift slightly. Beth Ferris had said that when she relaxed, her horse relaxed. I sagged like a pricked balloon and draped myself over Honey's shoulder like a preferred friend. I blew the soft exhalation of horse contentment. Honey's ears moved to their natural position, and she picked up her feet and crossed the ice.

How hard a horse's ears are pressed back tells the degree of its annoyance as well as its anxiety. Honey had been somewhat annoyed.

Pressed back hard against the head and twisted downward, ears express fury. The openings of the ears are pressed so tight during this movement that they close like a kid sticking his fingers in his ears to cut off the sounds that are provoking his anger. These are the ears a rider does not want to see, for fury means bucking and tossing the rider if possible.

The ears, eyes, and nose together can speak more subtly about anger. A gelding on the Beech Hill Farm approached by a burly Sunday rider laid back his ears and rolled his eyes downward, showing the whites. His nostrils flared. The message was that the horse was so extremely uncomfortable with the situation that an experienced handler would realize he was intent on biting the rider or bucking him if he tried to mount. This is not an uncommon horse statement at riding stables where almost every rider is new and strange. Like John Schrader, most handlers listen to this horse talk and find the irritating person another horse.

## HOOFBEATS

Beneath the horse are four wildly expressive instruments—legs and hooves. No one mistakes the meaning of a kick, but there is a gradation of intensity from front to rear.

"When Sunny strikes out with her front foot, she is certainly irritated with me," said Beth.

"But a kick with the hind foot, the cannonball of horse talk, warns me that she's so mad I'd better get out of the way immediately. I heed and talk softly to calm her before I approach again."

Beth reads her mare's gait as well.

"Sunny trots along stiff legged at times, a warning that she's feeling irked and will buck and kick. I have to yell at her to bring her out of it.

"And then there is the loose leg, placed gently on the ground. This is a beautiful message. It says you are doing everything right. It is usually followed by that wonderful sigh of the horse, air going out through the lips and nose, a message of pleasure and contentment."

Pawing expresses more frustration or impatience than

| THE HORSE'S TAIL POSITIONS | | |
|---|---|---|
| 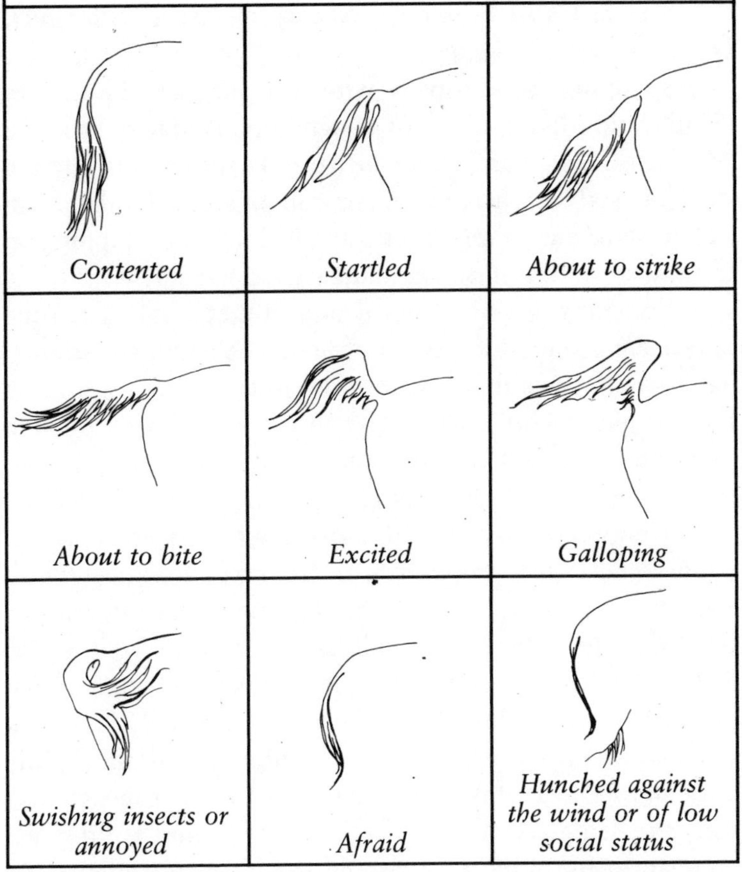 *Contented* | *Startled* | *About to strike* |
| *About to bite* | *Excited* | *Galloping* |
| *Swishing insects or annoyed* | *Afraid* | *Hunched against the wind or of low social status* |

downright anger. Often it is directed toward where the relief from frustration can be expected, as when Wauwinit pawed toward Rosie. Pawing can therefore easily be conditioned. Circus horses are taught to paw for rewards of treats and pats.

Exaggerated pawing, with one leg raised high or, on occasion, both front legs lifted, escalates a request to a threat: "Better do what I want, or I am going to strike."

The feet also beat out auditory messages—*stamping, knocking,* and *thumping.*

Stamping, done usually with just one front foot, is an identification like a name or a signature. Audacity stamped "Audacity. Here I am" when he heard Diane coming into the stable. In stables where the horses cannot see each other from stall to stall, they stamp frequently, like prisoners tapping on cell walls, to keep in touch and ease the loneliness.

Knocking is a hind-foot drumroll that results from the movement required to get rid of flies. The auditory meaning has become generalized to include other pests. A knocking horse hopes to rid itself of any insect, dog, horse, or man that annoys it. Beth's horse Sunny knocked when I was introduced to her; she has not changed her opinion of me.

Thumping, a more rhythmical sound than either stamping or knocking, is produced with front feet, back feet, or all four. It is a protest. A mare coming into estrous thumps to tell the stallion she is not ready and to keep back. It is also a protest when directed at people. A horse will thump if it does not want to be ridden. The ears are usually laid back to make the protest perfectly clear. A horse will thump lightly with all four feet while standing in place when, eager to take off, it is being restrained. The hoof expression is often accompanied by biting or gnashing the bit in impatience. As Zane Grey put it, "The mustang snorted and champed the bit, ready to bolt."

## ONE EVENING'S CONVERSATION

One evening I listened with pleasure to the clop-clop of Wauwinit's feet as Rosie brought him into the barn for the night. It is a moody sound of a weary horse ending the workday. I had heard it every evening on my great-grandfather's farm

in Pennsylvania, a taps to the fields and meadows.

Now, as my education proceeded, my perceptions of the horses' language had improved under the tutelage of Rosie and the big walnut stallion. Wauwinit's well-paced but fairly heavy sound was obviously that of a horse being led to shelter for the night, but it was not so weary sounding that I could be sure he was really done with the day. In fact there were a few things left undone. Rosie translated his talk as they entered the barn.

Several yards along the corridor between stalls Wauwinit stopped. His nostrils dilated, his head and neck were extended. His mouth was closed, one front foot was thrust out.

"He smells something close by that interests him," Rosie explained.

She answered his curiosity in English.

"It's a new barn kitty, Wau-Wau." His nostrils had already given him the message. He now saw it and looked at it intently.

"Horses are enrmously curious. They'll stop and look at anything that smells or looks new. Even a rock on the trail." That made sense to me. How could a horse make a fast getaway or avoid a new hazard unless curiosity piqued it to learn of each change in its environment? Wauwinit's ears rotated as he lifted his head and walked on. His eyes and ears were oriented to the side.

"All's well?" I guessed out loud. Rosie nodded. I really was learning horse talk.

As we neared his stall, Furrow, a gelding who is retired after winning many races, stamped. He was making a statement of his presence, probably to Rosie. Another horse knocked as we walked by his stall.

"That's for Wau-Wau." Rosie was sure. "That horse wants Wau-Wau far away. Hear him say so?" She glanced into the dark stall at a beautiful bronze-colored Thoroughbred.

"His name is Helpful Henry."

Wauwinit laid his ears back.

"Look at his eyes," Rosie said. "They are telling Henry

that he's feeling aggressive. They are so wide open that they seem a size too large.

"Also look at his nostrils. They are dilated and drawn back so far that they wrinkle his nose. Very aggressive. He doesn't like Henry, and he is getting madder the longer we hang around here." She tried to hustle him along but had to wait while Wauwinit declared his manhood to the gelding by arching his neck and giving the stallion pose for the barn community. He nickered. Henry was scared and screamed.

Safely past Henry's stall, I asked Rosie if Wauwinit had ever threatened her that way.

"Once, severely." She shook her head. "I was in the stall with him and the door was open when Henry came by. Wau-Wau wanted to get to him, but I was in his way.

"He turned, opened his mouth, and showed me his teeth. That's the worst threat. But if you want to show you're dominant, and not get bitten, you step forward, not back. I went right up to him, raised my arm, and scolded—loudly. He backed up and calmed down, but it was touch and go for an instant."

We continued slowly down the long corridor between the stalls where eleven horses are stabled. We passed a gelding who had been indoors all day waiting to be shod by the blacksmith. His head was low, almost between his legs. The corners of his mouth were drawn back.

"He's lonely," Rosie remarked, then reached over the stall and clicked her fingers. "Come, Summer Afternoon." She glanced at me. "Horse names are incredible. The namers get them out of books, newspapers, all kinds of places." She turned back to the horse. "It's okay. We're all here now." But his face remained unchanged. Horses certainly have nothing resembling a smile.

Wauwinit was led into his stall, where Rosie removed

his halter. She went on with her other duties. An hour later I saw Wauwinit's ears had drifted to face out to the sides, the rest position. His neck, relaxed, was nearly horizontal, the lower lip was drooping, and his eyes were beginning to close.

"That's horse for 'I'm going to sleep.' " Rosie translated automatically, although no other interpretation seemed possible. She checked the door, the lock, and the halter hanging by Wauwinit's name and walked across the concrete floor and down the ramp leading out into the yard. The sound of her boots echoed through the warm horsey-smelling barn. Her day, a twelve-hour routine from 5 A.M. to 5 P.M., was over. We stood a moment at the barn door. I heard only the sound of horses breathing as they slept, standing on all fours, ready to run as their ancestors ran if there were any disturbance in the night.

"Good-night," Rosie called.

Wauwinit stamped his answer. I heard him sigh, "All's well."

## VOICED COMMENTS

Ask any child what a horse says, and he or she will whinny. Strangely, the vocal sounds of horses, their most expressive communications in our perception, are rather fuzzy in meaning unless understood in context. Thus a stallion screams to challenge, but Helpful Henry screamed in fear. George H. Waring, a horse behaviorist, identifies four vocal horse sounds: the *squeal,* the *nicker,* the *whinny,* and the *groan.*

Squeals are voices for a big-sky country. They can be heard for a quarter of a mile. The open-mouthed, high-pitched outcry *eeeee he he he* expresses great emotional arousal; seeing an antagonist, the horse uses it as a defense threat to handler and other horses. It is a fighting word. The head is pulled in toward the neck and is moved from side to side while the horse squeals. Wauwinit squealed when he ran the hilltop and saw another stallion. He squealed when he was feeling combative, or if he was threatened or terrified. A mare squeals when she does not want the stallion's advances and before lactating if she is touched on the sensitive mammary glands.

The nicker is a low-pitched, guttural sound that seems to come from inside the head and is made with the mouth closed and nostrils wide open. It has several meanings. Every afternoon at the Birch Hill Farm the horses nicker at feeding time.

"It's an expression of anticipation," Schrader said. "Here in the stable it means 'The food is coming.' " He thrust his pitchfork into a bale of hay and lifted it over his head so that Jiggs could see it. The stallion nickered.

"These horses nicker if they just see one of the handlers at the right time of day, the late afternoon," he went on. "They don't need to see the hay or grain, just the handler. They have a sense of time."

A stallion nickers in anticipation if a receptive mare is nearby, and a mare nickers to her foal in anticipation of any danger she sees coming.

The whinny, which is similar to the nicker, begins with a slight squeal and terminates in a nickerlike sound. It is long and quite audible, and it is a statement of recognition to horse and person: "Hello, I know you." The mouth, as in the nicker, is closed; the ears and eyes are directed toward the recognized one; the nostrils dilate; and, as the horse whinnies its recognition, it walks toward its friend, as did Audacity toward Diane.

Curiosity arouses a whinny in a horse as if to alert others to a new barn cat, a strange object on the trail, a new person arriving in the barn.

The whinny is also a call to locate other horses when

## WHINNY CHART

*Each horse sound has several meanings. Notice context and watch posture to discriminate among them.*

| Sound | Meaning |
| --- | --- |
| Squeal: *e-he-eeeeee* (high-pitched, carries more than 100 meters) | A challenge or a defensive response to a challenge |
| | A mare's rejection of a stallion's sexual advances |
| Nicker: *uhn-uhn-uhn* (low-pitched and soft) | Anticipation of being fed or anticipation of the approach of a horse or person |
| | A stallion's anticipation as he approaches a sexually receptive mare |
| | A mare's call to her foal when danger threatens |
| Whinny: *e-he-eee-uhn-uhn-uhn* (begins as a squeal, ends as a nicker) | Recognition of an individual horse or person |
| | Curiosity, as in "What's going on?' |
| | Call to the band when separated from it |

separated. A mare whinnies to find its foal, and the foal to find its mother. In the wild the whinny of location brings wandering herd members back together again.

Horses, like humans, groan in pain—a low humlike sound that speaks of illness and discomfort and is usually answered by an owner with a call to the vet.

A slightly different groan, called the *sigh-moan,* speaks of psychological pain, usually the pain of boredom. The sound is a short groan followed by an exhalation through the nostrils.

"When I hear that sigh-moan," Diane told me, "I take Audacity out of his stall and ride him or groom him."

| | |
|---|---|
| Groan: *urghhhhh* (humlike) | Grunt uttered before lying down |
| | Repeated plaint during prolonged discomfort or boredom |
| | A remark that there is a suspicious object several feet away |
| Blow: *phruuuu* (uttered through wide-open nostrils) | An alarm to alert nearby horses |
| | Clearing of the nose after olfactory investigations |
| Snort: *phrugh!* (uttered through fluttering nostrils) | Complaint about being restrained by an obstacle, handler, or another horse |
| | An effort to rid the nose of dust |
| | An insult, as in "You're as irritating as a noseful of dust" |
| Snore: *zzzzzzzzz* (raspy, similar to a human snore) | The preamble to blowing an alarm |
| | The labored breathing of a recumbent horse |
| Snore: *brzzz, brzzz, brzzz* (like a chain saw) | The horse is asleep |

## TOUCH AND GO

Grooming and riding, the two things people spend most of their time doing with their horse, are also the most intimate expressions of their relationship. They are touching, and touching is cozy among all mammals.

A horse grooms itself with the tongue on the spots it can reach—the forelegs and shoulders. But its stiff backbone prevents it from cleaning other parts of its body. Instead, horses keep clean by grooming one another. Standing head to tail, with shoulders touching and tails switching to mutually fend off flies, the pair sniff from the crest of the neck to the withers, the shoulder, and along the back to the base of the tail. This same route is then nibbled between the teeth. The grooming is more than just practical, for it expresses affection and bonds individuals with one another. In the wild, grooming is so personal that it is usually done only among members of the same social unit—mothers and offspring, sibling and sibling, or best friends. Because of his many admirers' attentions, the lead stallion is usually polished to a glow, making him, like his counterparts in other animal societies, a glistening and gorgeous beast.

Although handlers and owners groom for good looks and healthy skin, the contact also binds man and horse. When Audacity was suffering from a skin problem and was feeling depressed, Diane walked him into the corridor of the barn and groomed him while waiting for the vet to arrive. She was as concerned about the horse as he was depressed about himself, but the contact helped both their spirits. As the grooming progressed, Audacity held his head higher and higher, and his tail was out from between his legs by the time the vet arrived. Asked how she felt after grooming the horse, Diane smiled. "Better."

Finally, riding is the ultimate language between man and horse. Nonriders, and, unfortunately, many ill-trained riders, too, think that a horse is made to go as a machine is made to go. Pull the rein, as one would pull the handle on a sled, and the horse turns. Accelerate with a kick; brake with a pull. The fact is that a horse can be guided perfectly with no gear at all, as the Indians who rode without bit, bridle, saddle, or stirrup knew.

"The trick," said Cliquo, the riding teacher at the camp in Maine I attended as an adolescent, "is to make your body at one with the horse's body, but moving in slight anticipation of what movement you wish next of the horse."

"I want to go forward." I told her.

"Shift your weight forward. As a horse moves forward, it shifts its weight forward. Predict that motion for your horse. It will follow through!" I leaned—magically the horse walked forward.

"How do I stop?"

"Shift your weight to the rear—lean back!" I followed her advice, and the horse, having been told through my weight to shift its own weight backward, slowed and halted.

Turning is the same idea. The horse compensates for your weight shift to the left, which throws it off balance, by turning left.

"Think of steering a horse like steering a bicycle with no hands," Cliquo would call from the center of the ring, her riding crop lifted like a baton. "It's the same principle."

Heels to the belly and tension on the bit make the message clearer.

"If I poke a finger under your ribs," Cliquo said, demonstrating on me, "what happens?" I pulled my hips away from her and bent to the right, demonstrating clearly why a horse that has been spurred in the right groin moves its hindquarters to the left, away from the thrust, and walks to the right. At the same time, tension on the right side of the bit in the horse's mouth makes it want to turn its head that way to relieve the pressure. I tested Cliquo's premises. The horse walked in a circle clockwise.

**CENTAURS**

Cliquo's instructions are easy enough to understand, but very hard to do. For those, the centaurs of this world, who fuse their body with their mount's body, anything seems possible.

The rapport between person and horse came as an astonishment to most of the viewers of the 1984 Olympic steeplechases, when they learned that the full-speed-ahead horses had never before seen the jumps, moats, turns, and hills. Their riders had walked the course, then through their bodies had anticipated for their mounts what was to come.

With that perfect communication the great running machine is back on the grasslands, its legs reaching out to leap, as its ancestors leaped, over brush and streams, its hooves pounding out the miles. We ride the captive horse, not just to get somewhere on legs faster than our own, but to set the horse free, and thereby ourselves.

The gallop may be wings for mankind, but the walk pace is fast enough for me. Beth Ferris called one particular walk with Sunny a poem.

"I was riding up the Rattlesnake Valley. It was a lovely afternoon in spring. Sunny was quiet. Her ears, attentively forward most of the time, flicked back occasionally toward me, listening, wanting to know what I wanted, how I was doing. But part of this perfect ride was that I was not concentrating too much on my horse—not too concerned about her. There is a happy partnership in walking, and we left each other pretty much alone to our thoughts. Hers ran, I suppose, to the green grass sprouting now after a long winter. Mine were on the clump of morel mushrooms I hoped to find growing under the cottonwoods near the creek."

That's all there is to the poem, the quintessential haiku of the horseman.

Two parts of the same body don't, after all, have to be concerned with one another as long as everything is going well. Perhaps that explains why, when horse and rider have been so accustomed to sharing the same body that they no longer think much about one another, a horse can act so intelligently when something goes wrong. A friend of mine who had polio as a child and cannot walk, rides because on her horse she can not only walk but run again. One day when she was out riding with a friend, the horse stumbled and fell with her. As he recovered, she slipped, saddle and all, to the side of her horse, where she dangled from her safety belt. As though the horse's own body had suddenly failed to function properly, he simply stood still. He did not move a muscle until her companion had unclipped her safety belt and lowered her safely to the ground. If that horse had not come to feel its human was physically part of it, it would have been scared to death and have bolted.

**TALKING THROUGH THE NOSE**

I should know.

Welcome, the horse who took me by his own precipitous route down the Grand Teton, was another bale of hay. In retrospect I can now appreciate that he was one of the best horse communicators I have ever encountered and that, though I thought at the time he was a horse mute, he had actually said a lot to me every day of the trip.

Welcome belonged to Ted Hartgrave of Moose, Wyoming. My children and I had planned a three-day pack trip up Death Canyon in the Tetons into the alpine meadows above timberline. Because I was detained with last-minute details, by the time I got to the corral, Welcome was the only animal not mounted.

As I approached him, his nostrils dilated and he went into a bout of sniffing that, had I been more experienced, I would have known was not the normal happy greeting sniff of the horse, for it was followed by the *blow*—air exhaled loudly through those expanded nostrils as though the horse had smelled something mighty unpleasant. "Something strange and frightening is here" might be a polite translation, but I think "This stinks" is more like it. As usual among horses, the message instantly went the rounds of that domestic group.

"Watch out!" All heads went up; ears rotated forward. A silence followed that, I now know, is predictable after a blow. The group focused on this unpleasant and possibly dangerous intrusion. I was off to a good start.

The blow was followed by the *rear-end threat*. That was well within my novice vocabulary, and I would have walked up the mountain on my own two feet had not Ted Hartgrave grabbed Welcome's bridle and held him securely while I mounted.

Released from Ted's grasp, Welcome thumped out with all four feet his testy comment on human slowpokes. Then Welcome snorted his impatience to be on his way. The *snort* is different from the blow—more forceful and shorter, with the air exhaled energetically through fluttering nostrils. A horse also snorts when it has a noseful of dust, and that exhalation is believed to be the sound's root meaning: "Remove the irritant." One cannot run with dust in the nose. One also cannot ride an irritated horse.

I answered the snort by clicking my tongue "Go." He understood this cricket sound, promptly took off at a run, and would have passed all the horses but for the fact that the first horse he encountered in the single-file line put him in his place with a shoulder bunt.

Welcome *snapped*—the colt's deference to his superior. Not knowing at the time what snapping was all about, I missed the nearly inaudible tongue clacking that surely followed. It was just as well. I also didn't know that snapping is done mostly by young animals under two years old, and knowledge of Welcome's youth might have filled me with doubts about the maturity of his judgment.

Instead, heartened that my mount had been put in his place, I squared myself away on his back with force and confidence. My body message got through, and for the rest of the morning we plodded quietly up the shadowy canyon at the end of the line.

At noon, horsemen young and old dismounted. Hartgrave showed the children how to untack the horses and turn them into the meadow. The kids and Ted joined us for lunch on a great monolith that hung out over the canyon and the valley. The land below spread out like a relief map.

From behind me, very quietly, a *snore*. This is not the snore of a sleeping horse, which sounds like a chain saw, but a purposeful sharp inhalation of air, similar to the rasping of air across the palate in a human snore. It is a sort of "Hey, look!" One can see what there is to look at by following the horse's gaze.

Welcome had my attention and that of two other horses. Their ears were pitched forward. All eyes were focused on a fat marmot sitting on a rock, his front paws clasped to his chest. Welcome gave the blow of suspicion and alarm, and the marmot whistled his alarm and disappeared into a crevice.

Thinking of my father's hunting dog, I said, "Good spotting," to my horse, took a twig out of his mane, and patted his neck. But he did not whinny as the other horses did when they received special attention from their riders. Welcome would not admit I was alive.

He and I settled into a truce over the next three days, and although he did not blow when I approached to mount after the first day, he never did like me. He gave me the noisy tail switch, pesky-fly talk, every time I put my foot in the stirrup.

On the last day I destroyed what fragile contact I had with him when, unbeknownst to me, I walked him up to the dominant horse my son was riding. The mare put back her ears. Welcome tossed his head upward and to one side in distress at being forced to approach an aggressive animal who was ready and willing to attack him. I, of course, did not read the message and pushed him on. With that, Welcome tucked his tail and rump, flexed his hind legs, and down the mountain we went in the uncontrolled expression of lighting out from danger. No "Whoas!" or pulls on the reins could stop Welcome for one-half mile of terrifying descent. Just below timberline he stopped, and together we recovered while waiting for the pack team. For the next two hours we walked slowly back to the corral, both of us on foot.

From Welcome I had heard and seen almost every non-vocal communication of the horse—the blow, the snort, the snore, the tail switch, and the snap. If ever there had been a test of human-horse communication, I had failed it.

I dismounted, packed up our gear, and, after getting the kids into the car, went back to say good-bye to Welcome.

How do you say good-bye to a horse? I asked myself. With a sniff and a whinny as in the greeting? Nose bump? I had seen it done, but it was not for me. Perhaps you cannot say good-bye in horse talk, for the herd stays together in the wild where their language evolved without the need for such a communication. Perhaps you just say, "I'm a friend. The sun is going down."

I walked up to Welcome—the blow. I hesitated, then opened the door to his stall. His head was back, and the whites of his eyes were showing. Summing up what I had learned from more experienced horse people during the trip, I cautiously approached that nonaggression zone at the shoulder, slipped my arm over his neck, and sighed. I drooped my lower lip, my own invention.

"I am a friend. The sun is going down."

I could see the nostrils expand for the blow, could feel the rear feet bring the rump around for the kick—but the ears did not go back, and the eyes did not widen.

Softly through his nose came a warm breath of air from deep in his chest. The sigh. Welcome's head lowered toward the horizontal, his eyelids closed ever so slightly, his mouth relaxed.

"I am going to sleep. All's well."

# Closer Than We Thought

At a barbecue in the suburbs of Athens, Georgia, I found myself seated on a spacious wooden deck among a group of psychologists and anthropologists, professors and their graduate students. As the night crossed over into morning, the philosopher in each emerged and the truth of yesterday became today's error.

We were discussing animal behavior as a route toward understanding nonhuman communication, and the conversation had been meandering like an old river.

"The first thing we have to do," said an authoritative voice to my left, "is to settle what the difference is between man and the animals so that we know who we are and who they are."

I searched my memory, recalling that when I was in college the word was that the difference between man and beast was our ability to make and use tools. We throw stones, spear food on forks, and carry things in jars and baskets. These achievements, experts believed then, were unique to *Homo sapiens,* and textbooks referred to our species as the *tool user.* That was the late 1930s and early 1940s.

I certainly was not going to mention this long-disproved notion to the esteemed researchers around me, some of whom had just heard the latest in nonhuman tool use from chimpanzee watcher Jane Goodall. While traveling in Serengeti National Park in northern Tanzania, she and her husband, Hugo van Lawick, brought their car to a sudden halt in an open savanna where two Egyptian vultures—white birds with golden cheeks and about the size of ravens—were standing before an ostrich egg. One of the birds had a stone in its beak. It threw back its head, flipped its neck forward, and hurled the stone. It struck the huge thick-shelled egg with a thud. The vulture walked up to the egg, saw that it had not broken, and, doing some obvious thinking, picked up the stone for another try. The next pitch missed the egg entirely, but with the third effort the shell cracked, and it took only a few more hurls to open it.

Sea otters also use stones as hammers. They collect large hard-shelled abalones in their teeth during dives to the sea bottom, then each picks up a stone with its forepaws. Surfacing, the otters roll onto their backs, place the abalones on their bellies, and, clasping a stone in both front feet, smash the shells open.

One of the Galápagos Islands finches uses a cactus spine or a sharp twig as a fork. First, it tests the tool's strength by pressing it against a limb. If it breaks, it seeks a stronger one. Then the bird transfers it to a foot while it uses its beak to drill into the wood of a decaying tree. When an insect is found, the finch takes the spine in its beak again and guides it into the hole. With precision it spears the insect and pulls it out. Shifting the spine to its foot, the bird then eats its meal from the pick.

Jane Goodall was the first to report that at least some animals—the chimpanzees she was observing at Gombe Game Preserve—make the tools they use. To get drinking water trapped in tree cavities, chimpanzees crumple leaves to make a sponge. They "fish" for termites with a probe they fashion by carefully stripping the leaves from a slender branch. The probe is stuck into the termite corridor. Considering themselves invaded, the termites grab the foreign intruder in their jaws and hang on while the chimp pulls them out and eats them like shish kebab.

Impressive as these technologies are, it could be argued that they are instinctive, programmed behavior not comparable to the thoughtful way even a toddler might use a stone one time to crush a peanut, another time to scratch a table, and still another to bang against a pot. That's what I had thought, until Crowbar, our pet crow, taught me a lesson.

Crowbar, who enjoyed playing with my children, was in the sandbox with them one morning. Presently my daughter came running to tell me she did not want to play with Crowbar anymore because he did not play fair. He took all the toys and hoarded them in the crotches of the apple tree. I suggested she slide down the slide.

"Crows can't slide," I explained of my plot to foil the bird. "The bottoms of their feet are horny to keep them from slipping off tree limbs. Crowbar will stick at the top."

Fifteen minutes later I looked out to see my daughter and son sliding down the slide and Crowbar at the top—stuck. He was forced to hop down. I felt smug about having outwitted a crow.

A moment later, Crowbar was back in the sandbox. There he picked up a plastic coffee-can lid, flew with it to the top of the slide, put it down, stepped in it, and—zoom.

The young woman sitting next to me at the barbecue apparently had been mentally going over similar territory. Perhaps, she wondered aloud, an individual animal's failure to pass creative invention along to others marks the boundary between human and nonhuman. No sooner were the words out of her mouth than she retracted them. She had remembered the macaques that have been protected for generations at the Monkey Research Center, Takasakiyama, Japan. For years the monkeys had accepted wheat, soybeans, and sweet potatoes tossed on the ground to supplement the foods they could forage for themselves. These treats, though eagerly sought, were unpleasantly gritty.

In a moment of thoughtful inspiration, a young female monkey scooped up a handful of food and ran to the stream. Seating herself beside a pool, she dumped her sandy ration into the water. The dirt sank to the bottom, the food floated to the surface, and she scooped up the clean morsels and ate them grit-free.

Her mother and siblings observed, understood what she was doing, and learned to wash grain, too.

Ultimately, the food-washing invention was taken up by the younger set. Conservative elders never did go for the new fad, but by the following generation, most had adopted food washing. It had become a tradition.

If that was not sufficiently startling—primates are certainly smart, after all—the spread of an invention among a group of birds in England was enough to make the hair stand on end.

Blue tits, little birds like our chickadees, learned to flip the lids off milk bottles and sip cream during World War II in the village of Swathling, England. The lid-flipping technique spread from a few birds to almost the entire town population of tits and, a year later, to populations in towns ten miles away. The milk industry answered complaints by urging customers to put stones on their porches to be placed on top of the bottles by the milk deliverers. The stones were put out, but customers found them beside, not on top of, the bottles in the morning. More calls to the milk company. Officials insisted the deliverers had put the stones on the bottles. One suspicious housewife got up before dawn and peeked out from behind the blind to see if this were the case. It was true. The deliverer had dutifully placed a stone on each bottle. She was perplexed. Then came the blue tits. Without hesitation they alighted on the stones and pecked at their edges until they shifted over and toppled to the ground. With a whack the pretty blue-and-gold birds flipped the cardboard lids and sipped cream.

The next ploy was to cover the bottles with towels. This halted the assault on the cream for several days, and then towels were found cast aside, and cream sipping resumed. A check at dawn revealed that birds in groups of three or four each took a corner of the towel, lifted it off the bottles while flying, and dropped it; then they flitted back to the bottles, flipped the lids, and sipped.

Diabolically, the milk deliverers placed the morning's delivery in a box open on one side, but barely taller than a milk bottle so that there was no clearance for the birds. This briefly foiled the blue tits, which were now accustomed to matching their wits with humans'.

But before long the blue tits, now on their mental toes, simply flew into the open-backed milk trucks while the milkmen shoved bottles into the box on the stoop, and sipped cream there.

My mind drifted on the wings of birds as the balmy night drew on and the conversation flowed—I was remembering a chicken.

Her name was Helen, and she belonged to my childhood

friend Virginia. She was one of those Easter gifts who grew from a cute fuzzy plaything into a serious personality. Helen was much more than a hen. She not only followed my friend but called her to meals by clucking furiously if Virginia did not hear her mother's dinner call. She conveniently laid an egg a day on a towel in the corner near the kitchen stove. She made a hobby of carrying pebbles from one bowl to another, for no apparent reason but that it amused everyone, not least herself. When Virginia and I played jump rope, Helen flew back and forth over the low swinging rope, clucking and carrying on in her "Look-what-I-did" egg-laying voice.

Then there was Sam. I would never have believed in Sam had I not met a professional naturalist who knew him and Peter, a boy of ten, and reported her observations to me. Sam was Peter's box turtle.

"Do you love me, Sam?" the boy asked as he gently held him against his throat beneath his chin. Slowly Sam opened his box, slowly extended his head and neck, and moved them gently back and forth against Peter's throat.

"He's saying 'I love you,' " Peter explained.

"Tell me you love me good," he insisted and tipped his chin against Sam. The turtle stroked Peter again.

"Let's go for a walk, Sam," Peter said, and slowly Sam pulled into his shell. Peter put him in his pocket and strode out the door and down the path.

I thought of all manner of scientific explanations for Sam's behavior: that the warmth of the boy's skin had made the turtle stretch his neck; that the pressure on his shell had provoked him to—something. But Peter's mother reported this:

"Sam sleeps with Peter—right on his chest."

"Of course Peter puts him there?"

"Oh, no. Sam climbs up the tumbled blankets when

Peter is asleep, walks onto his chest, and stops. He pulls his head and neck into his box. I presume he's sleeping too. He's very quiet."

Could it be that animals so different from ourselves can be brought out of their faraway worlds into communication with humans?

"Well, one thing is certain," a voice interrupted my musings. "We can't say that the difference between man and nonhuman animals is that only we use language."

"That's still controversial," said a graduate student in psychology whose mentor was one of the firm believers that animals cannot use language.

Most linguists concur that two criteria must be met for a system of communication to constitute a language. One is that the words or signs be symbols for something and recognized as such by the user. That may or may not be a reasonable demand. Jean Piaget, the Swiss psychologist who did more to unravel the thought processes of children than perhaps anyone, discovered that to a three-year-old, words are the objects they stand for. They use symbols without knowing that they do.

The other criterion for a communication system to be a language is that the symbols be combinable with one another to form novel phrases or sentences that are nonetheless understandable by others. Any three-year-old can do that.

Research efforts to resolve the question of whether or not animals could use language foundered for a while on the fact that even our closest relative, the chimpanzee, hasn't the vocal apparatus to speak. Chimps raised like babies within a human family had learned to grunt a few words like "cup" or "milk," but no more.

Then, in the late 1960s, psychologists began to circumvent that obstacle by using nonvocal languages. David Premack of the University of California at Santa Barbara used plastic symbols of different shapes to represent words. He claimed that chimp Sarah learned some 130 words and phrases with this technique, but admitted that she had created nothing new to say.

Psychologists Allen and Beatrice Gardner of the University of Nevada fueled the debate when they taught the chimp Washoe to communicate with humans in American Sign Language (Ameslan), a gestural language used by the deaf. Washoe astounded scientist and layman alike not only by learning 125 signs but by inventing a novel phrase of her own. One day, seeing a swan swimming on a lake, but not knowing the sign for it, Washoe excitedly signed out "water bird."

Skeptics Dr. Duane Rumbaugh of Georgia State University and his wife, Susan Savage Rumbaugh, suggested that the people who were teaching the chimps these novel nonvocal languages were so involved with the animals that they were suffering from the Clever Hans effect. Clever Hans, a horse who lived around the turn of the century, was credited for a time with being able to solve arithmetic problems. When shown an addition or subtraction problem written on a blackboard, he would stamp out the correct answer with his hoof. Close study showed that Clever Hans did not know the answers. Instead he was picking up subtle clues from his trainer, who was unaware himself of what he was doing, that told the horse when to start and when to stop stamping.

This made skeptics of almost all animal behaviorists for the next forty years.

To offset the possibility that researchers were unconsciously cueing their students, Rumbaugh invented a language based on ancient Chinese, dubbed it *Yerkish,* reduced the words to lexigrams, or symbols, and programmed them into a computer at the Yerkes Primate Research Center in Atlanta.

A special keyboard using the Yerkish lexigrams was designed, and the computer was connected to various appurtenances, such as a vending machine and a movie projector. In order to obtain, say, a candy, the chimp would have to ask for it in a certain way: "Please, machine, give [name of chimp] candy." Each word was a symbol over a button, and correctly stated requests were answered automatically. A bright young chimp named Lana was domiciled in a playroom fitted to a chimp's taste with swings, windows, ledges—and the computer. After a few demonstrations, Lana was adroitly pushing the correct lexigrams in the correct syntax to obtain all her needs: water, food, candy, movies of chimps. By pushing the right buttons she could call Tim, a graduate student and her human friend, to come play with her and relieve her boredom with the machine.

Lana toppled mankind's superiority complex when she not only learned which buttons to push to get what she wanted but forced her human friends to create new lexigrams for items that had not originally been programmed into the machine. One day she taught Tim to tickle her, a favorite chimpanzee pleasure, and then got the idea across to put the word "tickle" on the machine. After that, she could push the buttons to say, "Please, Tim, come tickle Lana." Over the years she learned 125 word-symbols, developed counting skills, and compared piles of objects to say which had the greater number.

"Twenty years ago," said Rumbaugh in the winter of 1984, "if you had told me that apes could learn words in a meaningful way, I and other scientists would have said, 'That's impossible.' Now that's no longer a question. The answer is yes, they do."

The naysaying graduate student was upholding the opposition. He referred to the work of psychologist Herbert Terrace of Columbia University and his little chimp Nim (full name Nim Chimpsky, a play on the name of linguist Noam Chomsky of M.I.T., a proponent of the idea that language ability is biologically unique to humans). Terrace and his students put Nim through forty-four months of intensive sign-language drill while treating him much as they would a child. He learned aptly, signing "dirty" when he wanted to use the potty or "drink" when he spotted someone sipping from a thermos. Nim, said Terrace, nonetheless did not master even the rudiments of grammar or sentence construction. Unlike children, his speech did not grow in complexity, nor did it show much spontaneity. Eighty-eight percent of the time, he "talked" only in response to questions from his teachers. Terrace studied the tapes of other apes and concluded, "The closer I looked, the more I regarded the many reported instances of language as elaborate tricks (by the apes) for obtaining rewards."

If so, an ape had certainly fooled me. I had visited the famous Lana in the company of Dr. Rumbaugh to see this wunderkind for myself. As we approached the two-way glass side of her playroom, the alert chimp turned, saw me with her friend, and began to show off. She swung down to her machine, which glowed with symbols for some fifty words at that time, and smiled a toothy chimpanzee smile. The symbols on the keyboard had translations behind the scenes for the professors and assistants, who could not remember them all.

Glancing at me once again, she began typing, moving her hands so rapidly that I could barely follow the sequence of buttons she pushed.

"Please, machine." She pushed two of the lexigrams. "Give Lana M & M's." Three more buttons were pushed. Tim, who had been taking notes behind a one-way window where Lana couldn't see him, suddenly stood up. Lana was looking at the "water" button. She wanted M & M's *and* water. She needed a new word. She needed "and" right then and there to circumvent the long procedure of pushing two full sentences of lexigrams.

"That's when the machine seems frustrating," Tim said to me later. "I can't give her the new word she wants fast enough. She thinks faster than we can work."

"I think it's a matter of cultivating a process that's inherent in chimps," Dr. Rumbaugh explained to me. "They can symbolize. They can learn; absolutely they can learn."

Dr. Rumbaugh knows, as every pet owner knows, that the more you talk to an animal, whether in its language or ours, the smarter and more aware it becomes.

"Language changes and enhances chimps' lives," he said. "The language-smart chimps are much more reflective and communicative. They interrelate with humans differently and more effectively and are more clever in learning and solving problems. Even their demeanor is changed. They are less destructive and much more careful in their interacting with people."

There is not an animal owner who would not agree. The cats, dogs, birds, and horses—even wild creatures that have been brought into a dialogue with humans—seem to grow more clever, considerate, and aware. They are more careful of us.

Or is it the other way around? It seemed to me, watching Lana spell out her wishes to "machine," that opening up an avenue of communication between our two separate species had increased our awareness, sensitivity, and intelligence as much as it had the chimps'. Perhaps we are becoming clever enough to see just how clever they have always been.

As I left the incredible Lana to manipulate her human friends with her machine, Tim accompanied me.

"At long last," he said enthusiastically, "we are beginning to talk to the animals. Soon I will be able to ask Lana the sixty-four-dollar question."

"And what is that?" I asked.

"What do you think of people?"

That question may have been answered by Koko, a hulking female gorilla taught sign language by Francine Patterson, a psychologist at Stanford University.

Koko has made her human companions aware not only that her breed is bright, but that it shares devious abilities commonly held to be unique to people.

Koko can lie, argue, and insult.

Writes Patterson:

> At six o'clock on a spring evening, I went to the trailer where Koko lives to put her to bed. I was greeted by Cathy Ransom, one of my assistants, who told me she and Koko had been arguing. The dispute began when Koko was shown a poster of herself. Using hands and fingers, Cathy had asked Koko, "What's this?"
>
> "Gorilla," signed Koko.
>
> "Who gorilla?" asked Cathy.
>
> "Bird," responded bratty Koko, and things went downhill from there.
>
> "You bird?" asked Cathy.
>
> "You," countered Koko.
>
> "Not me. You are bird," rejoined Cathy, mindful that "bird" can be an insult in Koko's lexicon.
>
> "Me gorilla," asserted Koko.
>
> "Who bird?" asked Cathy.
>
> "You nut," signed Koko.
>
> "You nut, not me," Cathy replied.

Finally, Koko gave up. Plaintively she signed, "Damn, me good," and walked away.

"What makes all this awesome," continues Patterson, "is that Koko, by all accepted concepts of animal and human nature, should not be able to do any of this. Traditionally, such behavior has been considered uniquely human; yet here is a language-using gorilla."

Koko can get into contrary moods, once considered the prerogative of humans. Patterson can almost program her actions when she is in such a mood. When she was breaking plastic spoons, an assistant signed, "Good, break them," and instantly Koko stopped breaking them and started kissing them. She knows she is misbehaving on these occasions and once described herself as a "stubborn devil."

Koko has a rich collection of insults that she has invented—"rotten stink" and "dirty toilet," as well as "bird" and "nut." These she applies to people whom she is not getting along with. During a fit of pique she referred to Patterson, whom she calls Penny, as "Penny toilet dirty devil"—a real achievement in creativity in any language.

Koko creates not only novel phrases and meanings but new symbols. She is terrified of alligators, although she has never seen a real one, just toothy facsimiles. For signing about them, she has invented her own word by snapping the two palms together in an imitation of an alligator's jaws closing. "Large alligator" is a big movement with her arms; "little alligator" is a tiny movement with her fingers.

Koko can refer to events removed in time and space, a sophisticated characteristic of human language known to linguists as *displacement*.

Patterson: "What did you do to Penny?"

Koko: "Bite." (The bite, which had occurred three days earlier, had been called a scratch by Koko at the time.)

Patterson: "You admit it?"
Koko: "Sorry bite scratch."
Patterson: "Why bite?"
Koko: "Because mad."
Patterson: "Why mad?"
Koko: "Don't know."

But was Koko really ignorant of her motive? A wonderful thing about language is that it can be used to deceive as well as inform, to evade as well as confront.

It was during a confrontation with a reporter that Koko might indirectly have answered the question, What do you think of people? The reporter wanted to know whether Koko, who had been raised among people her whole life, thought of herself as a fellow human. Translating, Patterson signed: "Are you an animal or a person?" Instantly came Koko's response: "Fine animal gorilla."

Until recently animal-speech teachers thought language had to be taught by humans to each individual primate and could not be passed on to other members of the species.

Again the animals proved us wrong. All the chimpanzees that learned sign and symbol language are now teaching their knowledge to offspring and to untutored adult chimps. All are conversing.

The conversation at the barbecue heated into an argument as to whether or not the chimps and Koko know what they are saying, and promptly stalemated. The question pivots on the word "know," for if Koko knows she is a "fine animal gorilla," then she is conscious, aware of herself and of her thoughts. Consciousness, particularly self-consciousness, has become the last bastion for those who would protect our uniqueness among animals.

All other differences have scattered like duckpins under strike after strike of facts.

Our esthetic sense?

The male bowerbird is an artist. With a sense of design he skillfully decorates his love arbor with colorful petals, berries, and "found" objects to please the critic—the female. Certain colors are more favored than others, and each ornament is placed to emphasize its position within an overall design. Moreover, one species of this bird, the regent bowerbird, paints with a brush and pigment. He mixes earth colors, plant pigments, and charcoal with saliva, dips a wadded leaf or piece of bark in the paint, and daubs the walls of the bower he has built. The results are murals of passionate gray-blue or green.

Our ability to plan, to imagine a future event?

The alpha male wolf can figure out a strategy for attacking prey and, what is more, communicate the plan to his hunters.

Wage war? Commit murder?

There are those who find our violence and inhumanity toward man the blatant difference between us and the animals. But chimps in the Gombe Game Preserve wage war on neighboring groups that infringe on their territory, and they kill members of their own group for motives no easier to discern than "senseless" human murders.

"Our knowledge of death," a young woman had concluded earlier in the evening.

"No," I spoke up.

I had heard a Pulitzer Prize–winning scientist speak to this difference one night in Arizona, stating that the knowledge of death was the important difference between man and beast.

But had he ever seen a crow find a dead companion and bell out the death knell? Had he ever seen a crow die of strychnine-poisoned grain and then heard the flock mournfully announce the death? And did the speaker know that the crows, seeing such a disaster, learn which corn is poisoned and tell their offspring and friends?

The question of consciousness is less easily laid to rest. Crows may fear poison as wolves fear hunters, and both may connect such dangers to the sense of loss death arouses in humans too. But do they think, I, too, will die? Or do even uncannily "human" responses like Koko's self-identification as a "fine animal gorilla" represent merely mechanical learning? She might, after all, have been showing no more comprehension of her words than a child who has been taught to recite the alphabet, but who is unaware of what the letters stand for or that they can be spelled into meanings.

The discussion on that suburban deck ended around three o'clock when the moon was setting behind the pines. As we arose to leave, I felt unsatisfied. After hours of hashing and rehashing we had not concluded anything. I approached our host, the most venerable and renowned member of that late-night group of philosophers.

"So what does it all mean?" I asked.

"That we are closer to the animals than we think," came the answer.

As I stepped down from the deck, I felt I was stepping down from a tower we had mistakenly built to keep us above all other animals. I felt my kinship with the singing bird, the howling wolf, the blowing horse, and the head-bumping cat. I could see myself in them and was amused as well as impressed. I went to my room thinking of old Will Cramer. It had seemed to me, when I began to research the how-to of talking to animals, that the Will Cramers of this world would supply the magic—the intuition that, as he had told me, was "beyond words"—and that scientists would supply the hard facts. Now I realized that was not quite the case. For all their careful study, researchers eventually come face-to-face with communication that is beyond words, and for all their innocence, gifted laymen eventually uncover undeniable facts.

Nele, Sara Stein's big furry dog could not be housebroken. Sara went to work every morning and returned around five. As she walked in the door, Nele, who had held her urine all day, purposefully emptied her bladder on the carpet. Unconstrained by the scientist's refusal to anthropomorphize, Sara thought like a mother. Nele was as frightened as a young child when her human mother vanished, and, childishly, she got back at her by peeing on the rug.

She remembered how she had helped her children overcome their insecurities when she first went to work by taking them to her office a few times so that they could picture where their mommy was when she was not at home.

On a Monday morning Sara put Nele in the car and drove her into New York City. She parked, and they rode the elevator to the sixth floor, where they got off and walked into Sara's office. Nele was tense and nervous, and Sara let her explore the several rooms and corridors of the office for about ten minutes before tying her to her desk.

At noon Sara got up from her desk, stretched, let Nele off her leash, and tried to interest her in a new toy she had brought for the occasion. She had thought to please the dog, who loved to play ball, with the additional excitement of a toy that had a squeaker. The ploy was a mistake. The new toy squeaked the cry of an injured pup and inhibited her bite. Nele tried to play but literally could not hold on to this crying thing. She dropped it and looked apologetically at Sara.

Then brightness came over Nele's face. She abruptly turned, trotted into the boss's office, and returned with a tennis ball. The ball had lain on the third shelf of a bookcase. It had not been pointed out to Nele when she went exploring;

indeed, no one had been in the room then. She had seen it just once, and briefly—and had remembered it. She brought it to Sara totally aware of what she was doing, and the two played ball until the lunch hour was over.

But what about the house training? Could the dog, like a child, comfort herself with an image, a mere thought, when the actual person was not there? Apparently so, for after three consecutive visits, Nele never again wet the rug.

We each make our own observations. We each—mystically like Will Cramer and his dog Nick or scientifically like Archibald and his whooping crane Tex—can toss ropes across the abyss.

As I finish writing this book I am training Qimmiq, my malamute, to stay on his leash during the day. Ours is a town with leash laws and a dogcatcher, and Qimmiq runs away and stays too long. I set him free only on walks in the woods. Around home he is leashed, and this is fine except for one matter: Qimmiq does not like to eliminate while on the leash. He cries to be turned free, promptly eliminates, and instantly sneaks off. Consequently I am determined that he remain leashed and learn to eliminate despite his distaste, just as other dogs must.

One morning I went out to change his water. He bored into my eyes, whimpered, then lowered his head and tail and unhappily shook the leash. His language was perfectly descriptive and understandable. Never had a dog asked more clearly to be let free.

"No!" I said. "The dogcatcher will get you. You wander, you stay too long . . . " I added a long string of objections. Qimmiq whimpered again. He was getting nowhere. I was resolved.

Then he changed his tactics. He spanked the ground to invite me to play, which I love to do and always do with the leash off. He was definitely thinking, and not only that, he was thinking with cunning. I laughed out loud as I understood his plan, and my resolve faded.

"Will you run away if I let you go?" He flashed me a look that took me by surprise. The eye contact was steady and long, the face smooth and unwrinkled, a confident beta saying to his alpha, "Trust me."

What could I do? I unsnapped the leash. He bounded into the bushes, relieved himself—and returned.

For me the abyss had closed.